MODERN
IGNEOUS PETROLOGY

Modern Igneous Petrology

MOHAN K. SOOD

Northeastern Illinois University
Chicago, Illinois

A WILEY-INTERSCIENCE PUBLICATION
JOHN WILEY & SONS,   New York • Chichester • Brisbane • Toronto

QE
461
.S62

Copyright © 1981 by John Wiley & Sons, Inc.

All rights reserved. Published simultaneously in Canada.

Reproduction or translation of any part of this work
beyond that permitted by Sections 107 or 108 of the
1976 United States Copyright Act without the permission
of the copyright owner is unlawful. Requests for
permission or further information should be addressed to
the Permissions Department, John Wiley & Sons, Inc.

Library of Congress Cataloging in Publication Data:
Sood, M. K.
  Modern igneous petrology.

  "A Wiley-Interscience publication."
  Includes bibliographical references and index.
  1. Rocks, Igneous. I. Title.

QE461.S62   552'.1   81-820
ISBN 0-471-08915-X   AACR2

Printed in the United States of America

10 9 8 7 6 5 4 3 2 1

To Minni, Sanjay, Rishi, Parm-Gun, and my father

Preface

Phase equilibrium data on synthetic silicate systems, which are compositionally analogous to igneous rocks, form an integral part of modern petrologic study and research. It is, therefore, desirable that students in geology develop a good comprehension of the application of such data to the crystallization of magmas. Students in my courses have frequently expressed and stressed the need for a condensed and concise text in this area with a petrologic viewpoint.

This book is aimed at fulfilling this need and can serve as a text for a one-semester sequence in igneous petrology for undergraduate and graduate students. The book will also be useful as a reference and source text for workers in the field, professional scientists in related disciplines, and practicing geologists. It includes a review of recent literature and extensive references. Though a knowledge of basic chemistry, mineralogy, and petrologic terminology is assumed, important and relevant terms are explained in footnotes throughout the text.

A concerted effort in the preparation of this book has been made to condense, organize, interrelate, and interpret the available but widely scattered information on phase relations in petrologically pertinent silicate systems with a view toward providing a physicochemical framework for explaining the crystallization behavior of melts and their direct correlation with the rock series. A particular focus is placed on the crystallization tendencies in basaltic and alkaline magmas.

The book includes most of the important, simple to complex, synthetic silicate systems made up of components representing the major rock-forming mineral groups (pyroxenes, olivines, alkali and plagioclase feldspars, silica, feldspathoids, and iron oxides) conforming to the bulk chemistry of the igneous rocks. The major features of systems are highlighted with the help of polythermal diagrams to facilitate understanding of crystallization paths and crystallization intervals. It is attempted to demonstrate a continuity of changes by subtraction of phases along the phase boundaries and at the invariant points to develop a lineage of possible liquid trends from parental compositions to the final end products. An overview of the igneous process and the state of the art of experimental petrology are presented in Chapter 1, followed by a discussion, in Chapter

2, of phase relations in systems related to basalts under anhydrous one atmosphere pressure conditions. Such conditions illustrate at or near-surface crystallization behavior of natural melts. The controls of original melt composition, equilibrium, and fractional crystallization; solid–liquid reaction relation on the liquid lines of descent; compatible–incompatible relations of minerals; and the partial pressure of oxygen during crystallization are also discussed. Chapter 3 focuses on systems compositionally analogous to the residual liquids derived from the fractional crystallization of basaltic magmas (or chemically equivalent synthetic systems), demonstrating continuity and interrelations among mafic and felsic magmas. The role of liquid immiscibility in magmas is also covered.

The effects of volatiles and volatile pressure on crystallization trends in magmas are discussed with the help of systems under water vapor and carbon dioxide pressures in Chapter 4. Attention is drawn to the depression in the melting and crystallization temperatures and the shifts in the compositions of invariant points and phase boundaries in synthetic systems with increasing volatile pressure and their possible role in nature. A special feature of Chapter 5 is a section on recent data on solubility of volatiles, particularly water and carbon dioxide, in synthetic and natural silicate melts (magmas) under various pressures and temperatures.

Melting relations of rocks under different physical conditions are presented in Chapter 6 to discuss and evaluate the behavior of the whole rock system in terms of temperatures of appearance and disappearance of mineral phases, crystallization sequences, crystallization intervals, and melting temperatures in relation to similarities of behavior in synthetic and natural systems. Such information also serves as a basis of physical parameters for magma formation by partial melting.

Magma generation in terms of source material, controls of the depth-degree of melting, depth of magma separation and volatiles on the composition of magmas and mechanisms of melting, and so on are investigated in Chapter 7 as a logical culmination to the discussion of crystallization in the magmatic process. Also included is an explanation of the compositional consistencies of voluminous basaltic flows. The depth-temperature-composition framework to a 400 km depth is presented in Chapter 8 along with petrological-plate tectonic implications.

A comprehensive bibliography appears at the end to facilitate consulting of the original works for more details. It is hoped that students will find in this

Preface

book what they need to know in terms of the physicochemical basis about the hows and whys of magmatic crystallization.

Work on this book was begun during sabbatical leave spent at the Mineralogical Institute, University of Mysore, India, and the author is grateful to Professor M. N. Viswanathiah, Director of the Institute, and to many other students and colleagues for the facilities provided and courtesies extended during the stay. The author feels greatly indebted to all the scientists whose work formulated fundamental concepts and many related facts, which serve as the basis for this book in petrology. The enormous amount of information in the field necessitated that I be somewhat selective. The comments of my colleagues and suggestions of my students have been immensely helpful in the improvement of the manuscript. I would like to acknowledge the review of the manuscript by my friends B. V. Govinda Rajulu, A. Mottana, J. M. Piotrowski, and A. W. Forslev. Sincere thanks are also due to the time and devotion of Judy Dobryman for typing of the final manuscript and to Janet Korbus for enormous dedication and patience in drafting the diagrams. Rich Sato and Robert Wagner also deserve appreciation for their help. The travel grant by Northeastern Illinois University, Chicago, is gratefully acknowledged.

Mohan K. Sood

Chicago, Illinois
May 1981

Contents

LIST OF ABBREVIATIONS   xvii
1   INTRODUCTION   1

2   ANHYDROUS SILICATE SYSTEMS RELEVANT TO THE CRYSTALLIZATION OF MAFIC (BASALTIC) MAGMAS   8

    1   The System Diopside–Anorthite, 8
        1.1   Lever Rule, 10
    2   The System Albite–Anorthite: The System of Plagioclase Feldspars, 10
    3   The System Diopside–Albite–Anorthite, 13
    4   The System Diopside–Albite–Anorthite–Ferrosilite, 15
    5   The System Forsterite–Silica, 17
    6   The System Forsterite–Anorthite–Silica, 21
        6.1   Crystallization Trends, 22
    7   The Essential Features of a Four-Component Representation, 30
        7.1   Flow Sheet with Univariant and Invariant Relations, 33
    8   The System Diopside–Forsterite–Albite–Anorthite, 34
        8.1   The Limiting Systems, 34
        8.2   Phase Relations in the Quaternary System, 37
        8.3   Crystallization Trends, 38

9 The System Diopside–Forsterite–Nepheline–Silica: The Simple Basalt Tetrahedron, 41

9.1 *Features of the Tetrahedron Di–Fo–Ne–Sil, 42*
9.2 *Phase Relations in the System Di–Fo–Ne–Sil, 45*
9.3 *Flow Sheet with Univariant and Invariant Relations, 49*
9.4 *Liquid Trends, 51*

10 The System Forsterite–Nepheline–Larnite–Silica: The Expanded Basalt Tetrahedron, 55

10.1 *Flow Sheet with Univariant and Invariant Relations, 55*
10.2 *Liquid Trends, 58*
10.3 *Plagioclase–Melilite Incompatibility, 65*

11 Selected Mafic Silicate Systems with Iron Oxides, 65

11.1 *The System Forsterite–Wollastonite–Iron Oxide–Silica, 66*
11.2 *The System $MgO–FeO–Fe_2O_3–SiO_2$, 69*
11.3 *The System Forsterite–Anorthite–Magnetite–Silica, 72*
11.4 *Silicate Systems with Chromium Oxide, 76*

3 ANHYDROUS SILICATE SYSTEMS RELATED TO THE CRYSTALLIZATION OF RESIDUAL ALKALI MAGMAS 77

1 Nature of the Residual Liquids, 77

2 The System Nepheline–Kalsilite–Silica: Petrogeny's Residua System, 77

2.1 *Crystallization Trends, 80*

3 The System Diopside–Nepheline–Kalsilite–Silica: The Alkali Rock Tetrahedron, 82

3.1 *Characteristics of the Alkali Rock Tetrahedron, 83*
3.2 *Phase Relations in the System Di–Ne–Ks–Sil, 86*
3.3 *Flow Sheet with Univariant and Invariant Relations, 88*
3.4 *Liquid Trends, 96*
  Nepheline (Sodic) Trend, 96
  Leucite (Potassic) Trend, 97
  Peralkaline Trend, 99

Contents

  3.5 The Leucite–Liquid Reaction and the Pseudoleucite Formation, 99

 4 The System Diopside–Forsterite–Nepheline–Albite–Leucite, 100

  4.1 Flow Sheet with Univariant and Invariant Relations, 101

  4.2 Liquid Trends, 105

  4.3 Generation of Alkaline Magmas, 106

 5 The System Diopside–Forsterite–Akermanite–Leucite, 107

 6 Selected Salic Silicate Systems with Iron Oxides, 107

  6.1 The System $Na_2O–Fe_2O_3–Al_2O_3–SiO_2$, 108

 7 Liquid Immiscibility in Silicate–Iron Oxide Systems, 113

4 SILICATE SYSTEMS WITH VOLATILES 115

 1 Effect of Volatiles, 115

 2 The System Diopside–Anorthite–Water, 117

 3 The System Albite–Anorthite–Water, 118

 4 The System Albite–Orthoclase–Water, 120

 5 The System Albite–Orthoclase–Silica–Water: The Granite System, 121

  5.1 General Features and Crystallization Trends, 122

 6 The System Anorthite–Albite–Orthoclase–Water: The System of Feldspars, 125

 7 The System Anorthite–Albite–Orthoclase–Silica–Water, 128

 8 The System Nepheline–Kalsilite–Silica–Water: The Silica Undersaturated Portion, 130

 9 The System Forsterite–Anorthite–Albite–Silica–Water, 134

 10 Other Selected Silicate–Water–Carbon Dioxide Systems, 136

  Diopside–Acmite–Nepheline–Albite, Diopside–Sanidine, Forsterite–Nepheline–Silica, Forsterite–Kalsilite–Silica, $K_2O–MgO–Al_2O_3–SiO_2–H_2O–CO_2$

5 THE SOLUBILITY OF WATER AND CARBON DIOXIDE IN SILICATE MELTS   140

1 The Solubility of Water in Silicate Melts, 140

2 The Solubility of Carbon Dioxide in Silicate Melts, 144

6 MELTING RELATIONS OF ROCKS: THE WHOLE ROCK SYSTEM   148

1 Melting Relations of Basalts with Water, 149

2 Melting Relations of Basalts without Water, 153

3 Melting Relations of Phonolites and Agpaitic Syenites and Nepheline Syenites, 158

    3.1 *Chemical Parameters vs Liquidus Temperatures, 163*

    3.2 *Relationship Between Melting Intervals, Volatile Contents, and Agpaitic Indices, 163*

    3.3 *Combined Effect of Controlled Oxygen and Water Vapor Pressure on the Melting of Undersaturated Alkaline Rocks, 167*

    3.4 *Melting Relations vs Field Relations, 168*

7 MAGMA GENERATION   169

1 Source Materials, 169

    1.1 *Garnet Peridotite Model, 172*

    1.2 *Pyrolite Model, 175*

2 Melting and Magma Formation, 180

    2.1 *Heat of Melting, 180*

    2.2 *Low Velocity Zone, 183*

    2.3 *Magma Ascent and Magma Separation, 184*

    2.4 *Depth-Degree of Melting and Magma Compositions, 186*

2.5 *Compositional Consistencies in Voluminous Lava Flows, 193*
  2.6 *Role of Volatiles, 196*
 3 Summary, 197

8 CONCLUSIONS — 198

REFERENCES — 205

AUTHOR INDEX — 229

SUBJECT INDEX — 235

SYSTEMS INDEX — 243

List of Abbreviations

Ac	Acmite	$NaFe^{+3}Si_2O_6$
Ak	Akermanite	$Ca_2MgSi_2O_7$
Ab	Albite	$NaAlSi_3O_8$
Alk-Fels	Alkali Feldspar	$(Na,K)AlSi_3O_8$
Amph	Amphibole	—
Anl	Analcite	$NaAlSi_2O_6 \cdot H_2O$
An	Anorthite	$CaAl_2Si_2O_8$
Ap	Apatite	$Ca5(PO_4)_3(F,OH,Cl)$
Ct	Calcite	$CaCO_3$
CaTs	Ca-Tschermak molecules	$CaAl_2SiO_6$
Cg	Carnegieite	$NaAlSiO_4$
Cr-Sp	Chrome Spinel (Chromite)	$FeCr_2O_4$
Cpx	Clinopyroxene	—
C	Corundum	Al_2O_3
Crist (Cr)	Cristobalite	SiO_2
Di	Diopside	$CaMgSi_2O_6$
En	Enstatite	$MgSiO_3$
Fa	Fayalite	Fe_2SiO_4
Fels (Fp)	Feldspar (Ternary)	—
Fs	Ferrosilite	$FeSiO_3$
Fo	Forsterite	Mg_2SiO_4
Ga	Garnet	$(Ca,Mg,Fe)_3Al_2Si_3O_{12}$
Geh	Gehlenite	$Ca_2Al_2SiO_7$
Gl	Glass	—
Gross	Grossularite	$Ca_3Al_2Si_3O_{12}$
Hd	Hedenbergite	$CaFeSi_2O_6$
Hem	Hematite	Fe_2O_3
Hyp	Hypersthene	$(MgFe)Si_2O_6$
Il	Ilmenite	$FeTiO_3$
Fe-Sp	Iron-Spinel (Magnetite)	$Fe^{2+}Fe_2^{3+}O_4$
Jd	Jadeite	$NaAlSi_2O_6$
Ks	Kalsilite	$KAlSiO_4$

List of Abbreviations

La	Larnite	Ca_2SiO_4
Lc	Leucite	$KAlSi_2O_6$
Liq	Liquid	—
Mt	Magnetite	$Fe^{2+}Fe_2^{3+}O_4$
Mel	Melilite	$(CaNa)_2(Mg,Fe,Al,Si)_3O_7$
Mer	Merwinite	$Ca_3Mg(SiO_4)_2$
Mo	Monticellite	$CaMgSiO_4$
Ne	Nepheline	$NaAlSiO_4$
Ol	Olivine	$(Mg,Fe)_2SiO_4$
Or	Orthoclase	$KAlSi_3O_8$
Opx	Orthopyroxene	$(Mg,Fe)SiO_3$
Ph	Phlogopite	$KMg_3(AlSi_3O_{10})(OH)_2$
Pig	Pigeonite	$(Mg, Fe, Ca)(Mg,Fe)[Si_2O_6]$
Pl	Plagioclase	$(Ca,Na)Al_2Si_2O_8$
K-Fels	Potassium Feldspar	$KAlSi_3O_8$
Pr	Protoenstatite	$MgSiO_3$
Py	Pyrope	$Mg_3Al_2Si_3O_{12}$
Px	Pyroxene	—
Qtz (Qz)	Quartz	SiO_2
Ra	Rankinite	$Ca_3Si_2O_7$
San	Sanidine	$KAlSi_3O_8$
Sil	Silica	SiO_2
Sod	Sodalite	$3NaAlSi_4 \cdot NaCl$
Sm	Sodamelilite	$NaCaAlSi_2O_7$
Nds (Ds)	Sodium disilicate	$Na_2Si_2O_5$
NS	Sodium silicate	Na_2SiO_3
Sph	Sphene	$CaTiSiO_5$
Sp	Spinel	$MgAl_2O_4$
ss	Solid solution	—
Trid (Tr)	Tridymite	SiO_2
wt. %	Weight percent	—
Wo	Wollastonite	$CaSiO_3$
Wu	Wustite	FeO

MODERN IGNEOUS PETROLOGY

CHAPTER 1
Introduction

Igneous process may essentially be regarded as a manifestation of high temperature (or high temperature–high pressure) phenomenon exhibited in the outpouring of molten rock material as *lava* or its consolidation at depth as *magma*.

In a physicochemical sense magma is a multicomponent system consisting of a complex silicate melt phase of anions, Si–O linkages, Al–Si–O complexes, and free cations with some solid phases as suspended crystals. A volatile (gas) phase, comprised mainly of H_2O and minor amounts of CO_2, HCl, HF, H_2S, SO_2, H_2BO_3, and so on (see Table I), is also generally present (Nordlie, 1971; MacDonald, 1972). Upon cooling such a melt will be, for the most part, governed by the principles of physical chemistry. Cooling and crystallization of magma, with accompanying differentiation, may result in the formation of many different igneous rocks.

A fundamental problem of igneous petrology is the diversity in mineral and chemical composition of igneous rocks. It is a major concern of the petrologist to seek the origin of this diversity. We know from petrography that this compositional variation is not of a random nature. Instead, minerals of igneous rocks show compatibility and incompatibility relations (Bowen, 1928; Schairer and Yoder, 1964), and rocks themselves demonstrate association tendencies that define definite rock series. The members of a series generally have gradational chemical and/or mineralogical characteristics that point to a common origin. The process of magmatic differentiation, in response to chemical and physical equilibrium, may play a dominant role in the formation of different members of the series.

Some of the major objectives of modern petrologic research are:

1 To construct "quantitative" physicochemical bases of magmatic processes.
2 To elucidate the nature and mechanisms of magma differentiation.
3 To predict crystallization behavior of magma.
4 To ascertain the nature of magma regimes.

Table I Volatile Contents of Selected Rocks and Lavas[a]

Type	H_2O (wt.%)	Cl (ppm)	S (ppm)
Ocean floor basalts			
Tholeiites	0.25	—	
Hawaiian tholeiite	0.50	—	
Alkali basalt	0.90	—	
Lava dredged from 4000-m depth at Kilauean Rift	0.45	—	
Olivine basalts, Surtsey	0.70	—	
Kilauean tholeiite, 1955–60 eruption	0.060–0.10	—	
Average basalt	0.91	60	
Average andesite	0.86		
Average phonolite	0.96	2300	
Average rhyolite	0.78	—	150–85
Average granite	0.47	200	—
Leucite basanites	—	4700	—
Tholeiitic glass	0.08	—	380
Nepheline syenites	0.5–0.7	360–1800	215–910
Average volatile content of lavas			
Early eruptions	1.0–2.5 wt. %		
Late eruptions	0.20–0.70 wt. %		
Estimated for basalt at depth	0.5 wt. %		

Composition of Volcanic Gases for Selected Volcanoes (in volume percent)

	Japan	Lassen Peak	Surtsey	Mt. Pelee	Mauna Loa	Kilauea	Nyiragongo
CO_2	0.36	2.1	4.6	10.1	5.3	21.2	40.9
CO	—	0.6	0.3	2.0	1.4	0.8	2.4
H_2	0.14	0.4	2.8	0.2	4.4	0.9	0.8
SO_2	0.009	0.01	4.1	—	—	11.5	4.4
S_2	—	0.9	—	0.5	0.2	0.7	—
SO_3	—	—	—	—	—	1.8	—
Cl_2	—	0.3	—	0.4	0.2	0.1	—

Introduction

Composition of Volcanic Gases for Selected Volcanoes (in volume percent)

	Japan	Lassen Peak	Surtsey	Mt. Pelee	Mauna Loa	Kilauea	Nyiragongo
F_2	—	1.5	—	3.3	0.0	0.0	—
HCl	0.051	—	0.6	—	—	—	0.6
H_2O	99.39	93.7	83.1	82.5	73.2	52.7	43.2

^aCompiled from data in literature and from Macdonald (1972).

5 To define depth-compositional relationships of magma generation.
6 To explain diversity in and origin of igneous rocks.
7 To deduce compositional regimes of the earth.

Useful information on some of these aspects is obtained through detailed geological and structural field relations and its interrelation with petrographical and chemical data on rocks. However, there is much need to closely define the pressure–temperature *(P–T)* and compositional controls of the igneous process. Therefore, high temperature (and high temperature–high pressure) phase equilibrium studies on synthetic and natural rock systems, under controlled pressure–temperature conditions, have gained impetus over the last 30–40 years. Such studies constitute a special branch of petrology called *experimental petrology*. Experimental investigations help in developing a better understanding and interpretation of the complexity of crystallization of magmas responding to changes in thermodynamic environments caused by cooling and/or shifts in pressure.

The bulk chemistry of igneous rocks can, for the most part, be represented by eight oxides*—SiO_2, Al_2O_3, CaO, MgO, FeO, Fe_2O_3, Na_2O, and K_2O (see Table II)—which account for more than 90 wt. % of the rock composition. The proportions of these oxides vary with the rock type. A system made up of these eight oxides will be an ideal petrological system to study but one that is difficult to represent spatially.

Therefore, in experimental petrology, pertinent to the problems of magma crystallization and magma genesis, two approaches are commonly used. In one,

*Trace elements and volatile components, though important, are ignored for the present.

Table II Average Chemical and Normative Composition of Selected Igneous Rocks[a]

	1	2	3	4	5	6	7	8	9	10	11	12	13[b]	14[c]
	Alkali Granite	Grano-Diorite	Quartz Diorite (Tonalite)	Andesite	Basalt (Tholeiitic)	Basalt (Alkali Olivine)	Peridotite	Nepheline Syenite	Olivine Nephelinite	Olivine Melilite Nephelinite	Leucitite	Olivine Leucitite	Garnet Peridotite Xenolith	Carbonatites
SiO_2	73.86	66.88	66.15	54.20	50.83	45.78	43.54	55.38	40.20	37.59	47.11	43.64	45.5	9.58–SiO_2
TiO_2	0.20	0.57	0.62	1.31	2.03	2.63	0.81	0.66	2.90	3.40	1.25	2.54	0.2	0.65–TiO_2
Al_2O_3	13.75	15.66	15.56	17.17	14.07	14.64	3.99	21.30	11.32	11.09	15.74	10.28	2.7	2.90–Al_2O_3
Fe_2O_3	0.78	1.33	1.36	3.48	2.88	3.16	2.51	2.42	4.87	5.18	4.54	5.11	0.3	4.33–Fe_2O_3
FeO	1.13	2.59	3.42	5.49	9.06	8.73	9.84	2.00	7.69	6.77	4.54	5.89	7.0	4.37–FeO
MnO	0.05	0.07	0.08	0.15	0.18	0.20	0.21	0.19	0.22	0.35	0.27	0.15	0.1	0.72–MnO
MgO	0.26	1.57	1.94	4.36	6.34	9.39	34.02	0.57	13.28	13.47	5.24	13.86	41.9	6.69–MgO
CaO	0.72	3.56	4.65	7.92	10.42	10.74	3.46	1.98	12.99	14.73	11.01	10.66	1.9	34.06–CaO
Na_2O	3.51	3.84	3.90	3.67	2.23	2.63	0.56	8.84	3.14	3.62	2.02	2.16	0.20	1.02–Na_2O
K_2O	5.13	3.07	1.42	1.11	0.82	0.95	0.05	5.34	1.44	1.33	6.72	4.09	0.10	1.47–K_2O
P_2O_5	0.14	0.21	0.21	0.28	0.23	0.39	0.05	0.19	0.78	1.02	0.44	0.63	0.03	0.40–BaO
H_2O	0.47	0.65	0.69	0.86	0.91	0.76	0.76	0.96	1.08	1.45	0.87	0.72	—	0.81–SrO
														29.29–CO_2
														1.86–P_2O_5
														0.08–(Nb,Ta)$_2O_5$
														0.73–F
														— –ZrO_2

	1	2	3	4	5	6	7	8	9	10	11	12
Qtz	32.2	21.9	24.1	5.7	3.5	—	—	—	—	—	—	—
Or	30.0	18.3	8.3	6.7	5.0	6.1	1.7	31.1	—	—	15.3	6.9
Ab	29.3	32.5	33.0	30.9	18.9	18.3	4.7	32.0	—	—	14.2	6.1
An	2.8	16.4	20.8	27.2	25.9	24.7	7.5	2.8	12.8	10.3	19.0	13.8
Lc	—	—	—	—	—	—	—	—	6.5	6.1	9.1	9.9
Ne	—	—	—	—	—	2.3	—	23.3	14.2	16.5	—	—
C	1.4	—	—	—	—	—	—	—	—	—	—	—
Wo	—	—	0.3	4.2	10.3	10.8	3.9	2.1	17.2	11.6	15.7	17.8
En-Hyp	0.6	3.9	4.9	10.9	15.8	7.1	14.8	1.2	13.1	9.4	11.4	14.5
Fs	1.1	2.9	4.1	5.3	11.2	2.9	2.6	0.8	2.2	0.8	2.8	1.1
La	—	—	—	—	—	—	—	—	1.6	8.8	—	—
Fo	—	—	—	—	—	11.5	49.1	0.1	14.1	17.0	1.2	14.1
Fa	—	—	—	—	—	5.0	9.6	0.1	2.7	1.7	0.3	1.2
Mt	1.2	1.9	2.1	5.1	4.2	4.6	3.7	3.5	7.2	7.7	6.5	7.4
Il	0.5	1.1	1.2	2.4	3.8	5.0	1.5	1.4	5.5	6.5	2.3	4.9
Ap	0.3	0.5	0.5	0.7	0.5	1.0	0.1	0.4	1.8	2.3	1.0	1.5
Hem	—	—	—	—	—	—	—	—	—	—	—	—
Ac	—	—	—	—	—	—	—	—	—	—	—	—

[a]Analyses 1–12 from Nockolds (1954).
[b]After Carswell and Dawson (1970).
[c]Hyndman (1972).

systems of only a few components (minerals and/or oxides) are studied under different temperature, pressure, and compositional environments. With an understanding of the phase chemistry of these simple systems complex natural systems can be more easily understood and explained. A kind of order to the scheme of things in terms of hows and whys can be established. Studies in silicate systems chemically analogous to magmas have greatly contributed to our knowledge of natural petrologic phenomena. Moreover, close correspondence between natural crystallization sequences and those derived from chemically equivalent silicate systems is suggestive of similarities in cooling behavior of natural and synthetic silicate melts.

In the other approach the complex system represented by the rock itself is studied under varied conditions, for example, temperature and pressure. By these means valuable information concerning crystallization sequences, crystallization intervals, and estimates of the magmatic temperatures can be obtained.

These approaches provide, in a rigorous physicochemical framework, fundamental information on the origin and crystallization of various rocks and a means to test the merit of hypotheses formulated through conventional petrological and geological studies. Furthermore, it enables prediction of magmatic behavior and derivation of pressure–temperature data for a quantitative model approach to magmatic evolution in petrology.

Phase equilibrium studies of synthetic and natural silicate systems have evolved tremendously since the pioneering work by James Hall and J. H. L. Vogt in the late eighteenth and nineteenth centuries, respectively. The establishment in 1906 of the Geophysical Laboratory at Carnegie Institution of Washington, D.C., was the important beginning for modern igneous petrology. Bowen, Andersen, Morey, and Greig were among the early workers at the Laboratory who initiated systematic investigations on various silicate systems, which have been greatly enlarged in recent years by Schairer, Yoder, Kushiro, and co-workers. The data on silicate systems have been crucial to many fundamental aspects of igneous petrology. Some of the important ones being Bowen's Reaction Principle, related to magmatic crystallization and differentiation; the parent–daughter concept in rocks; the nature of the residual liquids (Bowen, 1912, 1915b,c, 1922, 1928, 1935, 1945, 1947); and the crystallization of basalts (Schairer and Yoder, 1964; Schairer, 1967).

The refinement in experimental techniques and equipment and the achievement of technology to control temperature and water vapor pressure (P_{H_2O}) resulted in studying systems with water and other volatiles to assess the role of volatiles in magmatic processes and magmatic evolution. Such studies have

Introduction

produced classic works, for example, Tuttle and Bowen's (1958) study on granites, Yoder and Tilley's (1962) work on basalts, and Wyllie and Tuttle's (1959a, 1960a,b) investigations on carbonatites.

The development of technology to attain pressures and temperatures corresponding to the mantle has further enlarged the scope of studies in silicate systems and their application to petrology. The work of Yoder, Kushiro, and co-workers at the Geophysical Laboratory; Ringwood, Green, and co-workers at the Australian National University, Canberra; Wyllie and co-workers at the University of Chicago; and O'Hara and co-workers at Edinburgh University has greatly expanded our knowledge about the importance of the depth-degree of melting controls on the composition of basaltic magmas, basaltic and related magma generation by partial melting of the upper mantle material, the role of volatiles (H_2O and CO_2) in alkali basaltic and carbonatitic magma genesis, crystallization–differentiation behavior of basalts at high pressures, and the nature and composition of the upper mantle (Ringwood, 1975; Yoder, 1976; Green, 1973a, 1975; Wyllie, 1966, 1977a,b; O'Hara, 1965, 1968, 1971; Green and Ringwood, 1967a,b).

Osborn and co-workers have demonstrated the importance of partial pressure of oxygen (P_{O_2}), particularly in the formation of iron and silica enrichment trends during magmatic crystallization (Osborn, 1957, 1959, 1962, 1963). The ultrahigh pressure (in megabars*) and high temperature studies on silicate systems by Bell, Mao, and co-workers, solubility measurements of two or more volatiles in silicate melts at upper mantle pressures by Eggler, Mysen, Green, Wyllie, and co-workers, the determination of the extent of liquid immiscibility in silicate systems by Nakamura, Irvine, Naslund, and co-workers, and the study of melting relations of mafic-ultramafic rocks under mantle conditions and in the presence of volatiles by Green and other workers at the Geophysical and other laboratories are some of the recent innovations in experimental petrology.

The collective experimental work on silicate systems under many different pressure and temperature conditions by investigators all over the world is bringing us to the threshold of understanding the total igneous process and greatly enlarging our factual knowledge of the composition of the interior of the earth and the processes operative therein.

*1 megabar = 1×10^6 bars (for details of experimental techniques, see Mao and Bell, 1976).

CHAPTER 2
Anhydrous Silicate Systems Relevant to the Crystallization of Mafic (Basaltic) Magmas

The experimental investigations of anhydrous silicate systems at atmospheric pressure are applicable to, at or near surface crystallization of natural melts. We will subsequently confine our attention to the discussion of a series of silicate systems composed of end-member components of basalts at 1 atm pressure, to interpret the possible crystallization tendencies in mafic (basaltic) magmas.

Basalts are the dominant mafic volcanic rocks on earth. They are principally composed of almost equal amounts of plagioclase feldspars (labradoritic) and pyroxenes (diopsidic-augite) with minor olivine, nepheline, melilites, or hypersthene, and occasionally quartz. Although there are a few compositional variants, basalts show a limited range in mineral and chemical composition (see Tables II and XXI) in space and time. Therefore they are probably produced through melting of specific materials in specific pressure–temperature regimes in the earth, thereby retaining compositional continuities and similarities in many eruptions in many different geological environments of the world (Yoder and Tilley, 1962). They play a key role in the formation of many igneous rocks and may be "parental" magmas. Bowen (1928) had long advocated that fractional crystallization of basaltic magma could produce many different igneous rocks.

1 THE SYSTEM DIOPSIDE–ANORTHITE

In order to begin discussion of phase relations in systems related to basalts, it may be appropriate to consider first the binary system Di–An.* The system Di–An (Bowen, 1915a) is a very simple representation of basaltic composition,

---
*The various abbreviations for mineral names and related terms used in the text are given in the List of Abbreviations.

The System Diopside–Anorthite

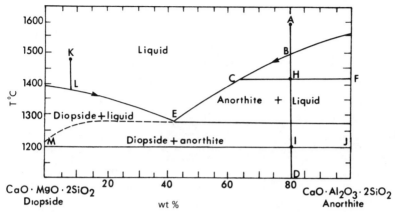

Figure 1 Phase equilibrium relations in the system diopside–anorthite. Note the solidus phase boundary is steeply curved, due to solid solution, for diopside rich compositions. (After Osborn, 1942. With permission of the *American Journal of Science*.)

as it contains both pyroxene and plagioclase components necessary to define basalts.

The liquidus surface of the system is shown in Fig. 1 and contains phase areas of An + Liq, Di + Liq, Liq, and solid Di + An. There is no solid solution* among the components. The system is a binary with a eutectic E. The temperature of the eutectic is 1274°C,† whereas its composition is $Di_{58}An_{42}$ (wt. %). The eutectic E is the lowest melting temperature and an invariant point where both diopside and anorthite crystallize simultaneously as the temperature and the composition remain constant.

Let us now consider crystallization of some compositions in this system. Upon cooling, a liquid of composition A begins crystallization of anorthite at the temperature of point B on the An + Liq phase boundary. As more anorthite crystallizes, the liquid composition moves down the phase boundary toward the invariant point E. At E the temperature and composition remain constant until the liquid completely crystallizes to a Di + An mixture in the proportion of the original composition A [liquid of composition K (Fig. 1), upon cooling, will

---

*The system is presented here as a simple eutectic; however, Clark et al. (1962) and Yoder (1965) have reported extensive solid solution in diopside, and the reader is referred to these works. Mineral phases in the silicate systems show a certain degree of solid solution because of their being portions of the larger oxide systems. For example, in systems containing Al compounds, diopside is generally somewhat aluminous. To avoid complexity, mineral phases are represented as simple components throughout the text.

†Most temperatures are within ±5°C of the stated values.

instead crystallize diopside as the first phase at point L and then follow the curve to E to give a final mixture of Di + An].

Only the liquid of composition E (Fig. 1) will begin simultaneous separation of Di + An with composition and temperature remaining constant, until the liquid completely crystallizes to a Di + An mixture in the proportion of the eutectic composition E.

The common ophitic texture in diabases-basalts, which is interpreted to represent almost simultaneous precipitation of pyroxene and plagioclase, indicates that basaltic compositions may closely correspond to the eutectic of this simple system.

The binary system Sph–An (see Prince, 1943) also has a eutectic relationship and should be studied for comparison.

1.1 Lever Rule

Let us say we want to determine the crystal–liquid proportion at point C (Fig. 1). In order to do this, draw a line AD projecting the original composition A to the composition axis. Now draw a tie line CF parallel to the composition axis and intersecting AD at H. It is then simple to determine the proportion of crystal and liquid phases, using the following relation:

$$\frac{CH}{CF} \times 100 = \text{Percent anorthite}$$

$$\frac{HF}{CF} \times 100 = \text{Percent liquid}$$

The lengths CH, CF, and HF can be measured from the diagram and substituted in the above relation. This is called the *Lever Rule* and can be applied to phase diagrams to determine the proportion of phases at any point during the cooling history of a composition.

Use the Lever Rule to determine the proportion of phases at point I in the lower part of Fig. 1.

2 THE SYSTEM ALBITE–ANORTHITE: THE SYSTEM OF PLAGIOCLASE FELDSPARS

As most rock-forming minerals show complete or partial solid solution, it is worthwhile to consider the relations in the system of plagioclase feldspars. This

The System Albite-Anorthite

will also be useful later to the discussion of the system Di–Ab–An, which is close to basalts in composition. The system Ab–An is an excellent example of complete solid solution. The phase relations are presented in Fig. 2, and the looplike arrangement of the liquidus and solidus curves is noteworthy.

A liquid represented by point A (Fig. 2) when cooled to point B precipitates plagioclase crystals of composition C, which can be read from the base. The important point is that the composition of the first-formed plagioclase is more calcic (An-rich) than the composition of the original liquid. Therefore the liquid becomes enriched in the sodic (or albitic) component. As the liquid further cools, the liquid composition follows the liquidus line to D, whereas the crystals change in composition by reaction with the liquid to the point E. At no point, under equilibrium conditions, could the liquid of one composition (e.g., point D) coexist with crystals of another composition such as point C (Fig. 2). In other words, liquid composition D can only coexist stably with crystals of composition E. Horizontal lines in Fig. 2 show stable liquid–crystal compositions. The liquid

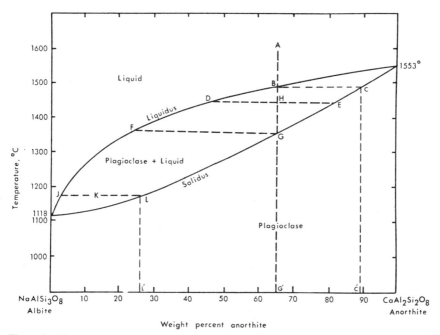

Figure 2 The system of plagioclase feldspars: albite–anorthite. (After Bowen, 1913. With permission of the *American Journal of Science*.) A zoned feldspar crystal with normal compositional zoning (core-An_{88}; Rim-An_{27}) will crystallize over a temperature interval of ~315°C.

will completely crystallize when crystals of the composition of that of the original liquid (point G in Fig. 2) precipitate.

The proportion of liquid to crystal could be determined by applying the Lever Rule. For a liquid of the composition D the amounts of crystal and liquid will be given as follows:

$$\frac{DH}{DE} \times 100 = \text{Percent feldspar crystals}$$

$$\frac{EH}{DE} \times 100 = \text{Percent liquid}$$

In other words, at H the lengths are proportional to the amount of solid and liquid phases.

However, if the reaction between liquid and early formed crystals is not complete—which is a likely phenomenon in nature, representing a condition of nonequilibrium and relatively rapid cooling—normal compositional zoning will result in feldspars, with Ca-rich cores and Na-rich rims. Zoned feldspars are frequently observed in rocks crystallizing under low pressures. Using the feldspar diagram (Fig. 2) and the compositions of the various zones, it may be possible to assess the temperature interval over which such rocks crystallized or obtain an upper limit of temperature of crystallization.

Fractional crystallization, the removal of early formed crystals, will undoubtedly (1) extend the total crystallization interval, (2) enrich the residual liquid in the sodic component, and (3) lower the temperature of final crystallization (Turner and Verhoogen, 1960). In general, very slow cooling (intrusive environments) will inhibit zoning, whereas in hypabyssal rocks zoning is more likely. In very rapidly cooled (or undercooled) lavas unzoned crystals of the same composition as the melt may form. Fluctuations in the physical and chemical conditions attendant at the time of crystallization may cause normal, oscillatory, or reverse zoning in the feldspars.

Yoder et al. (1957) have studied the phase relations up to a water vapor pressure of 5 kb. The temperatures of the liquidus and solidus are depressed by over 300°C, but the shapes of the liquidus and solidus are similar (see Fig. 36).

The system Fo–Fa is also an excellent example of complete solid solution, and the reader is advised to refer to Bowen and Schairer (1935) for comparative study. The systems Geh–Ak (Ferguson and Buddington, 1920) and Or–Ab (Schairer and Bowen, 1935; Bowen, 1937; Schairer, 1950; Bowen and Tuttle, 1950) are good variations, each representing a limited solid solution and a temperature minimum.

3 THE SYSTEM DIOPSIDE–ALBITE–ANORTHITE*

In order to more closely approximate the crystallization of basalts, let us now mix the components plagioclase feldspar and pyroxene. The resulting ternary system Di–Ab–An is certainly a better representation of a basaltic composition than the system Di–An.

The liquidus surface† of the system, from Bowen (1915a, 1928), is shown in Fig. 3. There are prominent phase areas of diopside (pyroxene) and plagioclase feldspars.

Some noteworthy observations about the system are as follows:

1. There is no ternary eutectic, however there is a boundary curve (the cotectic line *LMN*), which divides the plagioclase and pyroxene fields.
2. The lowering of crystallization temperatures occurs not only for a particular feldspar composition but also for final consolidation of the melt (cf. plagioclase system).
3. The crystallization of diopside is diminished as the liquids reach the cotectic line *LMN*. In turn, the residual melt is enriched in the sodic component.
4. The composition of plagioclase in the plagioclase field can be obtained by tie lines drawn through that point and projected onto the feldspar axis through the original composition for which the course of crystallization has already been determined experimentally.
5. The path of crystallization for each composition in the plagioclase field is curved, due to solid solution, and is unique for each different composition.
6. The collective phase relations and the solid solution nature of the components involved greatly enhance, through fractional crystallization, the variety of possible mineral assemblages (rocks) based on the proportion and composition of plagioclase feldspars. The composition and amount of plagioclase feldspar is one of the important ways of defining a rock type.
7. Fractional crystallization also increases the chance of the last liquid to reach or to terminate crystallization at the minimum.

---

*See Ehler (1972), Zavaritskii and Sobolev (1964), Hutchison (1974), and Levin et al. (1964) for discussion on various types of ternary systems, methods of calculation of proportion of phases, and graphical presentation of three-component systems. Schairer (1959) and Roedder (1959) are also good reviews.
†It implies melting relations of compositions in the system. The presence of a liquid phase is assumed unless otherwise stated.

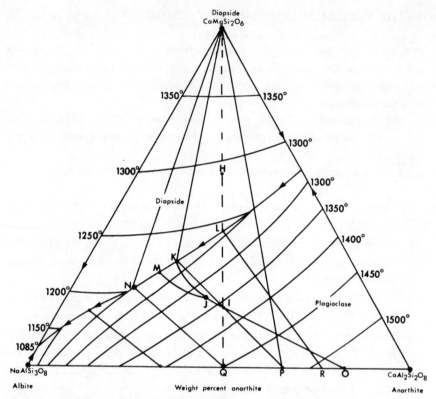

Figure 3 Equilibrium phase relations in the system diopside–albite–anorthite. The system of simple basalts. (After Bowen, 1915.) There is no eutectic or peritectic point, rather a cotectic phase boundary *LMN*. Liquid of composition *H* follows the path *HLMN*, while that of composition *I* follows the curved path *IJKMN*. The composition of last plagioclase crystals, under equilibrium crystallization, is given by point *Q*. The lines *LR*, *KP*, and *NQ* define coexisting diopside and plagioclase for the liquid composition *H*. Note that after reaching the cotectic, liquids *H* and *I* completely crystallize over a short temperature interval (<50°C) to two solids (Di and Pl) at some point *(N)* on the cotectic. (With permission of the *American Journal of Science*.)

8   Under equilibrium conditions crystallization will cease when plagioclase crystals of the original liquid composition separate.

9   The composition of the last plagioclase to crystallize for liquids in the diopside (or plagioclase) field can be obtained by drawing a line from the diopside apex through the original liquid composition (*I* or *H*) to intersect the Ab–An base (point *Q* in Fig. 3).

10   The distance the liquid will crystallize along the phase boundary *LMN* is dependent on the original composition and the extent of plagioclase–liquid equilibrium.

Upon cooling, the liquid compositions lying in either field will commence crystallization with a phase of the respective field (plagioclase will crystallize first for point I and pyroxene for point H in Fig. 3). With further cooling, the liquid will reach the boundary curve LMN and plagioclase and pyroxene will precipitate simultaneously as the liquid composition moves toward N. The liquid will finally crystallize when crystals of plagioclase of the composition of that of the original liquid precipitate. The final solid mixture will consist of Di + Pl.

It is important to note that a liquid of composition chemically equivalent to equal amounts of labradorite and diopside will plot close to the phase boundary and crystallize plagioclase and diopside almost together (cf. basalts) with residual liquid being enriched in the sodic component.

Let us consider crystallization for composition I (Fig. 3). At I (1350°C) crystals of plagioclase ($An_{75}Ab_{25}$) begin to crystallize. With further cooling the liquid follows the curved course IJK as the composition of plagioclase crystals moves from O to P. At K diopside begins to form and the liquid moves down the boundary KN, while the feldspar composition changes to Q. At this point crystallization ceases, as the plagioclase crystals now attain the composition of that of the original liquid.

However, if the crystal $\rightleftharpoons$ liquid equilibrium does not persist the feldspar may develop different forms of zoning: normal, oscillatory, or reverse, depending on the physical and chemical conditions attendant on the magma.

4 THE SYSTEM DIOPSIDE–ALBITE–ANORTHITE–FERROSILITE

Let us add a fourth component, ferrosilite, to the above system to satisfy the complex nature of pyroxenes. Figure 4 schematically shows the relations in the quaternary system Di–Ab–An–Fs. The boundary surface $ABCD$ in the tetrahedron separates the volume of plagioclase from that of pyroxenes.

In Fig. 4 point M represents a composition of ($Ab_{50}An_{50}$) = 60% and Di = 40%. Another point N on the join M–Fs will have the composition Fs = 20%, Di = 32%, and Pl = 48%, which is quite close to that of Deccan traps (see Table III). Thus this point N, representing a basaltic composition, will so plot on this diagram that only a very small amount of plagioclase or pyroxene needs to crystallize before it reaches the boundary surface where both will precipitate simultaneously (Bowen, 1928).

From the discussion of phase relations in the systems presented so far, it may be concluded that in the early stages of crystallization of basalts, pyroxene

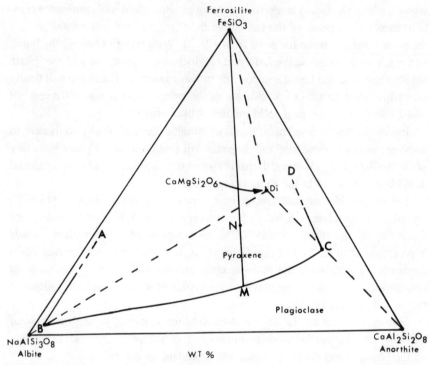

Figure 4 Schematic phase relations in the system diopside–albite–anorthite–ferrosilite illustrating possible similarities in the crystallization tendencies of synthetic and natural basalt compositions. (After Bowen, 1928. With permission of Princeton University Press.)

Table III   Composition of Liquid N and Average Deccan Trap[a]

Oxides	Liq. N (Fig. 4) (wt. %)	Avg. Deccan Trap (wt. %)
SiO_2	53.7	50.6
Al_2O_3	13.2	13.6
FeO	10.9	12.8
MgO	6.1	5.5
CaO	13.1	9.5
Na_2O	3.0	2.6
K_2O	0.0	0.7

[a] After Bowen (1928).

and plagioclase will tend to crystallize almost simultaneously. The frequent occurrence of ophitic texture in basaltic rocks supports such a deduction from synthetic silicate systems.

5 THE SYSTEM FORSTERITE–SILICA

So far we have not considered olivine as a component in the systems discussed. Olivine, however, is present in a great many basalts, and its presence generally defines a state of silica undersaturation. At low pressures olivine crystallizes in the early stages. If not separated, olivine undergoes Fo + Liq → En (pyroxene) reaction. The appearance of olivine in rocks is thus dependent on the silica saturation of the melt and the extent of crystal ⇌ liquid equilibrium.

The binary system Fo–Sil, studied by Bowen and Andersen (1914), is pertinent to the present discussion of systems related to mafic magmas. The liquidus surface of this system, which is a portion of the larger $MgO–SiO_2$ system, is shown in Fig. 5. It has an incongruently melting intermediate compound, enstatite,* which marks the correct silica saturation in the system. The phase diagram contains phase fields of Fo + Liq, Fo + En, En + Crist, En + Liq, Crist + Liq, and Liq. There are two invariant points, R and E. Point R is a peritectic for Fo + Liq → En reaction point, whereas E is an eutectic for En + Crist crystallization.

Let us now consider crystallization paths for some compositions in this system. Upon cooling, a melt of composition P (Fig. 5) begins to crystallize forsterite at Q. As the temperature falls, forsterite continues to crystallize, enriching the liquid in silica, until point R (1557°C) is reached. At R, temperature and composition remain constant as forsterite dissolves by reaction. It is important that for a composition such as P all forsterite is not changed over to pyroxene. The dissolution of forsterite ceases with the disappearance of the liquid phase. The final solid product therefore is a mixture of Fo + En.

For compositions between M and R (Fig. 5) forsterite is still the first phase to crystallize; however, all forsterite is made over to enstatite at point R, while some of the liquid still remains. With further cooling the liquid moves to E, at which point it is joined by separation of silica polymorph (Crist). The crystallization ceases as the last liquid is used up at this invariant point. The final solid phases are En + Crist/Trid.

---

*The term enstatite includes its polymorphic variants clinoenstatite and protoenstatite.

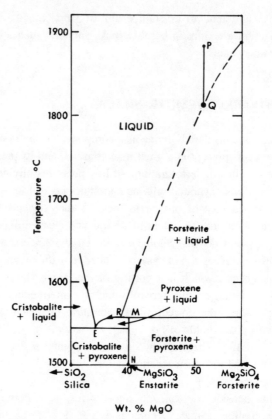

Figure 5 Phase equilibria in the forsterite–silica portion of the system MgO–SiO$_2$. (After Bowen and Andersen, 1914. With permission of the *American Journal of Science*.) Note entatite is an incongruently melting intermediate compound. The composition of the invariant *(peritectic and eutectic)* points is as follows:

Composition in wt. %

Invariant point in Fig. 5	Oxides		Fo	Sil	En	Sil
	MgO	SiO$_2$	Mg$_2$SiO$_4$	SiO$_2$	MgSiO$_3$	SiO$_2$
Peritectic R	39.0	61.0	68.25	31.75	97.5	2.5
Eutectic E	35.0	65.0	61.25	38.75	87.5	12.5

It should be noted that the line *MN* (Fig. 5), and its extension, is the line of silica saturation. All compositions to the right of this line are silica-undersaturated and will terminate crystallization at *R* to a solid Fo + En mixture, whereas those to the left are silica-oversaturated and will cease final precipitation at *E* to a solid En + Crist mixture. Only the composition on the line *MN* (or its extensions) can separate enstatite alone as the final solid phase. The above cases are possible if perfect equilibrium conditions prevail; however, in the case of supercooling, the Fo–Liq reaction may not reach completion. Therefore it may be possible to find an olivine and/or quartz (polymorph) association in rapidly cooled mafic melts, for example, quartz basalts or quartz diabases. The rate and degree of removal of olivine crystals can impart substantial flexibility to the process of fractional crystallization and greatly influence the course of crystallization of later liquids. Also small differences in the initial composition could cause substantial changes in subsequent crystallization.

The main implication here is that the Fo + Liq → En reaction does occur at low pressures. The extent of crystal-liquid equilibrium will endorse the derivation of the silica-oversaturation trend from the silica-undersaturated parent. The phase relations in this system provide explanation for certain important features observed in rocks:

1 Antipathy of olivine and quartz under normal cooling conditions (compatibility and incompatibility relation).
2 Presence of olivine and quartz in rapidly cooled mafic lavas.
3 Reaction rims of pyroxene around olivine.
4 Olivine tholeiite → tholeiite → quartz–tholeiite associations.

The addition of a third component, for example, diopside, anorthite, or albite does not inhibit the Fo–Liq reaction. It may, however, enlarge the temperature interval over which it takes place. Yoder and Tilley (1962) suggested that the extent of Fo–Liq reaction in the systems Di–Fo–Sil (Fig. 6*a*), An–Fo–Sil (Fig. 6*b*), and Ab–Fo–Sil (Fig. 6*c*) (or other systems containing forsterite and silica; Fig. 8) is dependent on the plot of the various compositions in the respective systems. They concluded that those plotting close to the Di–Fo join will show *no reaction,* while others will show *partial or complete reactions,* under equilibrium conditions. Subsequently the relations in one of these systems will be discussed.

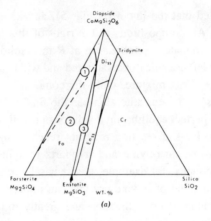
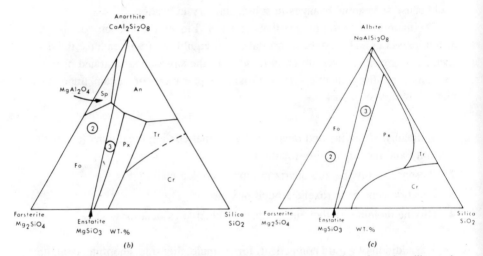

Figure 6 The extent of forsterite + liquid reaction in the systems containing forsterite, silica, and a third component (cf. Fig. 8a,b). The reaction relation of olivine in tholeiites. *(a)* Equilibrium phase relations in the system diopside–forsterite–silica. (After Bowen, 1914 and Osborn and Muan, 1960. With permission of the *American Journal of Science* and the American Ceramic Society.) *(b)* Equilibrium phase relations in the system anorthite–forsterite–silica. (After Anderson, 1915. With permission of the *American Journal of Science*.) *(c)* Equilibrium phase relations in the system albite–forsterite–silica. (After Schairer and Yoder, 1961. With permission of the Carnegie Institution of Washington.)

Areas marked by the encircled numerals 1, 2, and 3 indicate areas of none, partial, or complete Fo + Liq reaction, respectively. Olivine tholeiite compositions correspond to area 2, whereas tholeiite compositions correspond to area 3. Olivine fractionation will obviously lead to silica-saturated derivatives.

6 THE SYSTEM FORSTERITE–ANORTHITE–SILICA

It will be appropriate at this stage to combine the early and late crystallizing mineral phases to assess the compatibility relations and the nature of the residual liquids.

The three-component system Fo–An–Sil is a good example of such a system. Its liquidus relations,* studied by Andersen (1915), are shown in Fig. 7a,b. Certain important features of phase relations can be summarized as follows:

1. The system is a pseudoternary† due to the phase area of spinel.
2. The join En–An marks the limit of silica saturation.
3. The invariant point R (at 1260°C) is a peritectic for the Fo + Liq reaction, whereas the point E (at 1222°C) is a ternary eutectic. (T is a Sp + Liq reaction point.) Liquids in the triangle Fo–T–An (ATC in Fig. 7a) will leave the system and reenter at T after crystallization of spinel and will follow the line TR to the point R. This is also true of other systems containing the field of spinel but not spinel as a component.
4. The lowest melting composition is represented by point E, which is the goal‡ upon fractionation of most liquid compositions in the system.
5. Figures 7a,b show phase areas of forsterite, anorthite, enstatite, tridymite, cristobalite, and spinel.
6. The Fo–Liq reaction imparts, at shallow pressures, great flexibility to the crystallization-differentiation process and the extent of this reaction is a major determinant for the nature of the derivative liquids or magmas.

---

*Irvine (1975) has slightly revised the data on the system. He found the eutectic E at 1220°C, the peritectic R (Fig. 7a) at 1270°C, and the Fo–An cotectic TR to be curved toward the forsterite apex. The point T (Fig. 7a) is now at 1310°C rather than 1320°C and a little toward the silica apex.

†A system is pseudoternary if it contains field(s) of phases not represented by the components of the system, or if liquid composition(s) in part of the system move out of the system upon crystallization, as will be the case for compositions lying in the triangle ATC (Fig. 7a). All liquid compositions therefore, do not terminate crystallization in the system. Extensive solid solution may also be an important factor.

‡However, many melts may crystallize before reaching the eutectic. The extent of fractionation, solid solutions, and nature of initial bulk compositions may also control the path of the liquid and its approach to the lowest temperature point.

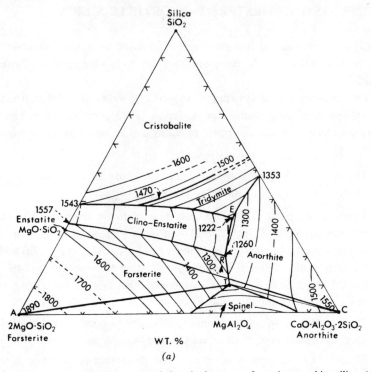

Figure 7 (a) Equilibrium liquidus phase relations in the system forsterite–anorthite–silica. (After Andersen, 1915. With permission of the *American Journal of Science*.) Note the join En–An marks the limit of silica saturation. The portion Fo–An–En characterizes olivine tholeiites and the portion En–An–Sil tholeiite compositions.

6.1 Crystallization Trends

We will now consider equilibrium crystallization* of three compositions K, M, and N lying on the silica-undersaturated, silica-saturated, and silica-oversaturated sides of the En–An join, respectively.

*Equilibrium crystallization implies perfect equilibrium being maintained between the liquid and crystals, that is, crystals may continuously react with the liquid, thereby changing composition, and remain in equilibrium with it as long as liquid is present.

In fractional crystallization perfect fractionation occurs, and crystals, as soon as formed, are effectively removed from the melt (magma) by various physical processes. The extent of reaction between crystal and liquid is minimal or none. This is the most effective process of magmatic differentiation (Bowen, 1928). However, a combination of both processes seems likely in nature, their dominance changing at different stages.

The System Forsterite-Anorthite-Silica

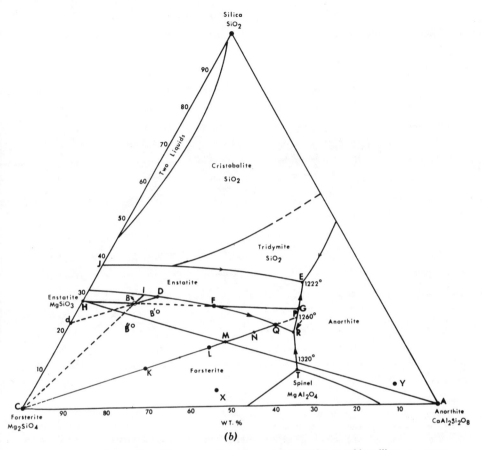

Figure 7 (b) Crystallization trends in the system forsterite–anorthite–silica.

A liquid of composition K on the silica-poor side of the En–An join (Fig. 7b) will, at appropriate temperature, begin crystallization of forsterite. With further cooling, the liquid will move along the line $KLMN$ directly away from the forsterite apex (because there is no solid solution) toward Q. At Q the liquid follows the curve QR to the point R. Along QR forsterite dissolves and enstatite crystallizes. At the invariant point R, temperature and composition remain constant, enstatite and anorthite crystallize, while more forsterite dissolves. The Fo–Liq reaction ceases as the liquid phase disappears. The final crystalline product thus is a mixture of $Fo + En + An$.

However, the composition such as M (Fig. 7b), lying on the En–An join and

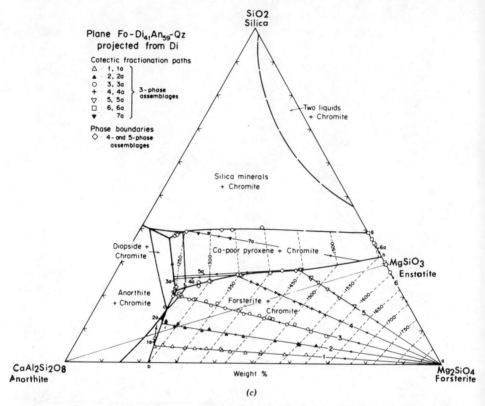

Figure 7 *(c)* Phase relations projected from diopside and picotite onto the plane forsterite–anorthite–silica. Numbers 1–6 refer to liquid paths on the forsterite-chromite cotectic. Note the expansion of chromite (spinel) field. (After Irvine, 1977. With permission of the Carnegie Institution of Washington.)

representing the stoichiometric silica saturation, will terminate crystallization at R, at which point anorthite also separates and all of the forsterite must be changed over to enstatite, by reaction, as the temperature and composition remain constant and the last of the liquid is used up. The final crystalline product is a mixture of *En + An and no forsterite*.

A liquid composition of point N (Fig. 7*b*) on the silica-rich side of the En–An join will also crystallize forsterite as the first phase; however, all the forsterite will react to form enstatite, while some liquid still remains. The residual liquid may now bypass R and leave the line QR to a point P. Along PE, En + An crystallize and the liquid composition moves toward E. At E, En + An + Trid

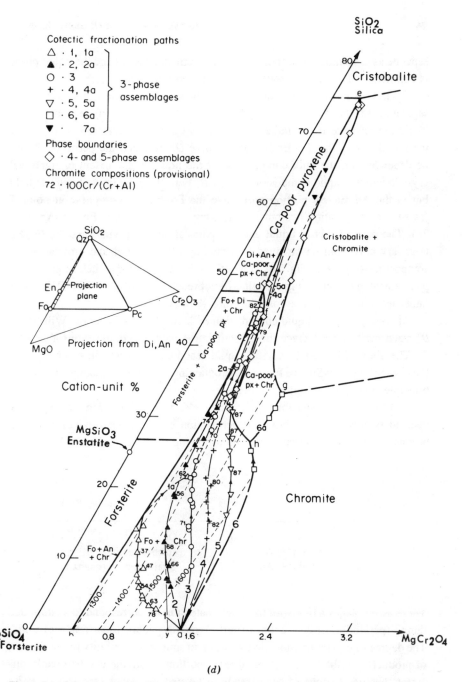

Figure 7 (d) Projection of liquidus relations from diopside of anorthite onto the plane forsterite–picotite–silica. This depicts phase relations contained in the tetrahedron as if projected from Di_{41} An_{59} apex. Fractionation paths are numbered 1 to 6 and show the 100 Cr/Cr + Al ratios. (After Irvine, 1977. With permission of the Carnegie Institution of Washington.)

separate as the temperature and composition remain constant and the liquid phase is eliminated. The final crystalline product therefore is *En + An + Trid*.

By now it must be evident from the above that the phase boundary *(QR)* separating the Fo–En (Fig. 7b) fields is a (Fo + Liq → En) reaction curve. Therefore enstatite is crystallizing as well as forming by reaction. The path of the liquids along this Fo–En boundary curve *QR* to reach the peritectic *R* will be dependent on the bulk composition of the liquid or the extent of fractional crystallization. Liquid composition (e.g., *B*, Fig. 7b) lying in the forsterite field but in the An–En–Sil portion will leave the Fo–En reaction curve at point *F* (as such a composition lies outside the compositional triangle En–Fo–An, Fig. 7b). The path of such a liquid and the point at which it will leave the Fo–En boundary curve can be obtained by drawing a line through the original liquid composition (e.g., *B* in Fig. 7b). If it cuts the Fo–Sil sideline at the compositional point enstatite (point *H*, Fig. 7b), it will not reach the peritectic point *R*, but will rather leave the curve to a point (*G*) directly across from it.

At what point will liquid composition *B'* leave the Fo–En curve? Will liquid *B''* reach peritectic *R?* Trace the crystallization path for compositions *X* and *Y*.

The above situation of crystal–liquid reaction also arises in other systems (Di–Lc–Sil, Fo–Lc–Sil; see Fig. 8a,b) containing reacting compounds, and these principles are equally applicable there.

In the cases discussed above, if points *L, M, N,* and so on (Fig. 7b) represent residual liquids obtained through fractionation of forsterite (olivine) from the original melt *K*, then we can have the following possible final assemblages:

Relevant Points in Fig. 7b	Assemblages	Representative Rock
L	Fo + En + An	Olivine gabbro
M	En + An	Gabbro (norite)
N	En + An + Trid	Quartz gabbro (quartz norite)

These assemblages along with the fractionated phase, correspond to a simplified dunite (peridotite)-olivine gabbro (norite)–quartz gabbro (quartz norite) trend. The degree of crystal fractionation will impart immense flexibility to the number of products possible since the relative proportion of olivine can be greatly different. Natural analogs of such trends in layered intrusions (see Irvine, 1970; Wager and Deer, 1953; Wager and Brown, 1957, 1968; Campbell, 1977; How-

The System Forsterite-Anorthite-Silica

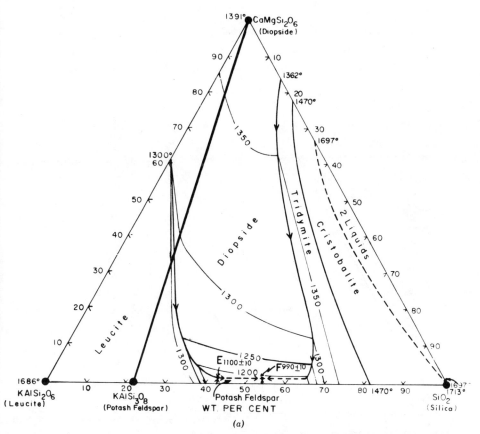

Figure 8 Phase relations in the systems containing reacting components. (a) The system diopside–leucite–silica with K-feldspars as an intermediate incongruently melting compound. (After Schairer and Bowen, 1938. With permission of the *American Journal of Science*.) The join Di–K-Fels marks the silica saturation. Liquid compositions to the left of this join finish crystallization at the peritectic E to a Di + Lc + K-Fels assemblage. Those to the right of this line are consumed at the eutectic F, giving a crystalline mixture of Di + K-Fels + Trid. Note that for many compositions containing about 2.0 wt % diopside, pyroxene is still the first phase to crystallize.

land et al., 1936; Daly, 1928; Lombaard, 1935; Hess, 1960) may represent the extent of olivine fractionation controlling the crystallization lineage of magmas of basaltic compositions. Only through fractionation can such liquids give silica-rich derivatives.

The above discussion leads to one very important observation. Very slight differences in compositions from that of En–An (Fig. 7b) join can substantially

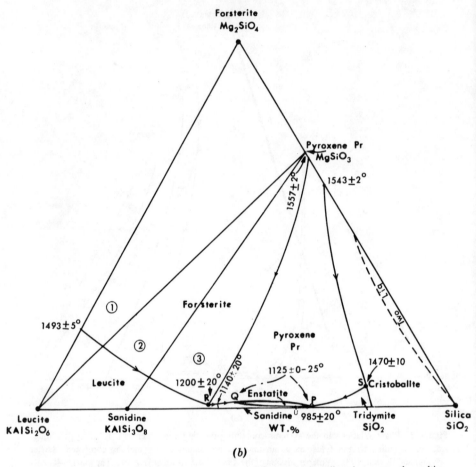

Figure 8 *(b)* The system forsterite–leucite–silica with two intermediate incongruently melting compounds, enstatite and K-feldspar. (After Schairer, 1954. With permission of the American Ceramic Society). Note large phase fields of forsterite and enstatite suggesting their early crystallization for many compositions containing even very small amounts (~1.0 wt %) of these components. Forsterite and leucite both show reaction relation with the liquid. Liquids such as 1, 2 and 3 will, under equilibrium crystallization, give Fo + Lc + En; En + Lc + K–Fels and En + K–Fels + Tr crystalline phases respectively. Fractional crystallization would produce a trend towards silica-oversaturation.

change the crystallization paths and the nature of the derivative liquids. This has vital impact on the nature and probable regimes of mafic magmas. Therefore magma types (and the resultant rock associations) may be controlled by depth of generation and degree of melting of the parental mantle materials. Compositional continuities among magma types may thus be possible.

In the system Fo–Di–Sil (Fig. 6a) an initially silica-poor melt can also yield, upon fractionation, quartz-bearing products. The magnesian pyroxenes at lower temperatures will continually react with the liquid to give hypersthene or pigeonite. The compositions close to the Di–Fo (e.g., 3) sideline will not undergo Fo–Liq reaction (Yoder and Tilley, 1962).

Compositions in the anorthite field have potentials of producing assemblages corresponding to troctolites and anorthosites (see also the system Di–Fo–Ab–An). Mixing of melts from near *TR* with those from *RE* (Fig. 7b) could give only anorthite as the liquidus phase and fractional crystallization would form the anorthositic assemblages (Irvine, 1975).

Irvine (1975) extended the studies in the system Fo–An–Sil with the addition of the orthoclase component to assess the role of salic contamination, magma mixing, and fractional crystallization mechanisms for the evaluation of the layered (stratiform) intrusions and, in particular, the explanation for the frequent presence of interlayered anorthosite association. Some of his observations are noted below:

1 Subalkaline-magnesian liquid compositions in the system Fo–An–Sil completely mix with potassic (granitic) liquids near the center of the Or–Sil join.
2 Liquids of chemically different compositions near their liquidus temperatures cannot mix adiabatically without a certain degree of crystallization.
3 Mixing of salic liquids with mafic (basic) melts favors crystallization of forsterite relative to anorthite and suppresses the separation of silica minerals.
4 A high degree of contamination of salic materials can take place in melts near the Fo–An cotectic line before olivine is eliminated as a primary phase. Compositions with as little as 6–7 wt. % normative forsterite have forsterite on the liquidus after 40–45% contamination. Therefore a high degree of cumulus olivine does not preclude salic contamination.
5 The intermediate melts obtained by mixing salic and basic compositions could produce anorthositic cumulates or pyroxene fractionation from such melts could move the liquid to the En–An cotectic line. The melt near the Fo–An cotectic, however, will be uncontaminated magma akin to the so-called "pulses" in layered intrusions. The data on the system Fo–An–Or–Sil clearly supports the association or interlayering of intrusions.
6 The crystallization trends in the system Fo–An–Or–Sil provide a basis for explaining the interlayered associations of anorthosites with orthopyroxenite, norite, troctolite, and peridotite in layered intrusions. Such interlayered as-

sociations of anorthosites have been described from many layered intrusions (see Cameron, 1963, 1971; Wager and Brown, 1975; Irvine and Smith, 1967; Wager, 1953; Van Zyl, 1970; Hess, 1960; Irvine, 1970).

7 ESSENTIAL FEATURES OF A FOUR-COMPONENT REPRESENTATION

Before proceeding to discuss some of the important quaternary systems, it may be appropriate to present a brief description of tetrahedral representation.

A regular tetrahedron may be used to depict phase relations in a condensed polythermal quaternary system. The essential features of the four-component representation have been described by Schairer (1942, 1954) and Bailey and Schairer (1966). A summary is presented here for convenience and clarification.

Figure 9a represents a simple condensed polythermal quaternary system A–B–C–D, where the four faces represent the four ternary systems A–B–C, B–C–D, C–D–A, and A–D–B. In each limiting system the primary phase fields are separated by three boundary curves (as AD-e_1, AC-e_1, and CD-e_1 in the face A–D–C), which intersect at a ternary invariant point (as e_1 in face A–D–C) at a specific temperature. At this point three solid phases $(A + D + C)$ are in equilibrium with a single liquid composition defined by that point (e_1 in the face A–D–C). Such a point can be a eutectic or peritectic depending on the nature of the solid phases and temperature distribution along the boundary curves. In quaternary systems univariant lines, trending into the volume of the tetrahedron, originate at such points.

Heterogeneous equilibria in a four-component system involving a liquid phase can be described with the aid of Fig. 9a.

1. The primary phase areas are interlocking volumes (volume of D shown as the shaded area, Fig. 9a). Each volume encloses all liquids in equilibrium with *one* common liquidus phase.
2. Any two adjacent primary phase volumes are separated by a curved divariant surface (e.g., I-e_2-BD-e_4 the surface between the volumes of B and D) defining liquids that are in equilibrium with two solid phases.
3. The contact between any three-phase volumes is a curved univariant line (e.g., line Ie_1) defining liquids in equilibrium with three solid phases. Such

Essential Features of a Four-Component Representation

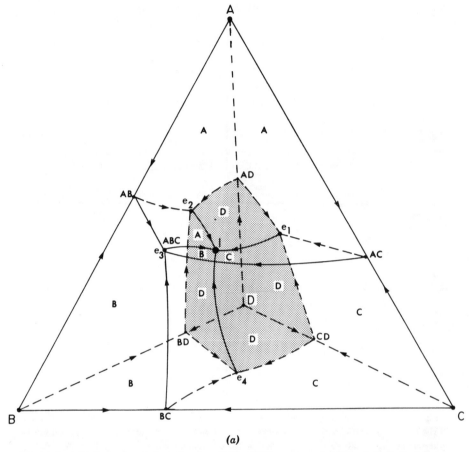

(a)

Figure 9 (a) Schematic relations in a hypothetical four-component system $A-B-C-D$. I is a quaternary invariant point. Note the phase volumes ($I-e_4-CD-e_1-AD-e_2-BD-e_4$ for phase volume of D) divariant surfaces ($I-e_2-BD-e_4$ for phases $B + D +$ Liq) and univariant lines (Ie_1, for $A + D + C +$ Liq). The arrows indicate direction of falling temperatures.

univariant lines originate at the peritectic or eutectic points in the limiting ternary systems or joins.

4 Four primary phase volumes can meet only at one point (e.g., I). This is an invariant point. Only a liquid of composition I may coexist with four solid phases at one temperature. Such a point represents the point of intersection of the four univariant lines.

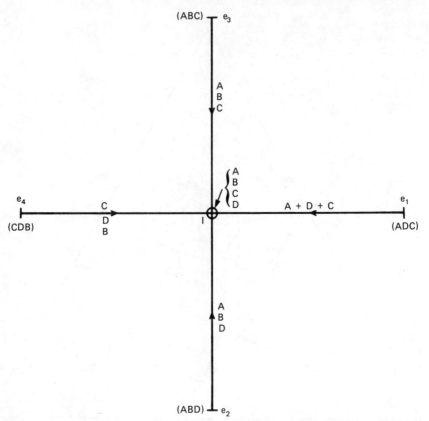

Figure 9 (b) Flow sheet showing the univariant and invariant relations in the system A–B–C–D. The four constituent ternary systems are given in parentheses. The lines e_1I, e_2I, e_3I, and e_4I are univariant lines. Shown are the three solid phases coexisting with liquid along these lines. Arrows indicate the direction of falling temperatures. At point I, a quaternary eutectic, $A + B + C + D$ coexist with liquid.

Experimental studies of a four-component system are principally concerned with the determination of the univariant and the invariant characteristics of the system. In such cases separate temperatures are not required, as the temperatures and liquid compositions may not vary independently.

The study of phase relations in the joins* (or planes), which are sections through the tetrahedron, help define the univariant and invariant phase relations

---

*If a join in a quaternary system intercepts a univariant line involving three phases whose compositions lie within the join, then the point of interception is not only a piercing point but also a peritectic or eutectic—a temperature maximum (Schairer, 1942). (See the joins Di–Fo–Ab, Di–Fo–Ne, Fo–Ab, and Di–Ab in Sections 7 and 8.)

Essential Features of a Four-Component Representation

in a quaternary system. They also divide the system into smaller four-component subsystems. Let us, for example, consider the compound "BC" to be an intermediate binary compound on the line B–C (Fig. 9a). Then the join A–BC–D will divide the system into two quaternary subsystems, A–B–BC–D and A–BC–C–D, each having its own quaternary invariant point. Two situations arise:

1 If BC is a congruently melting intermediate compound, then the join A–BC–D is a true compositional join, as its composition lies in the join (small amounts of solid solution will not drastically affect this). Thus the invariant points in the subsystems are both quaternary eutectics. Such a join is an effective "thermal divide" for equilibrium crystallization from either of its sides. Liquids will terminate crystallization at the eutectics in the respective subsystems.

2 However, if BC is an incongruently melting intermediate compound, then the quaternary invariant points in the subsystems A–B–BC–D and A–BC–C–D will be a peritectic and a eutectic, respectively. The join A–BC–D will not be a compositional join. The liquids will terminate crystallization at the quaternary eutectic point, chances of which are enhanced through fractional crystallization. Thus there are chances of obtaining separate lineages.

It is important and useful to determine the temperature and composition of the piercing points because these points serve as coordinates for univariant lines trending to an invariant point. Such data are used to construct flow sheets showing the univariant and invariant characteristics of the quaternary systems, for example, Di–Fo–Ne–Sil, Fo–Ne–La–Sil, Di–Ne–Ks–Sil, and Di–Fo–Ne–Ab–Lc discussed in later sections.

7.1 Flow Sheet with Univariant and Invariant Relations

Flow sheets, though schematic, show the paths of liquids to the various quaternary invariant points and low temperature points for the termination of crystallization. Such liquid paths define possible rock series. The arrows on the univariant lines always indicate the direction of falling temperatures.* The divergent arrows define planes that are thermal divides.

---

*As the temperature on univariant lines trending to an invariant point must fall in a four-component system without solid solution, the lowest temperature on a univariant line defines the maximum temperature of the invariant point. Moreover, the temperature along the univariant lines may increase if the solid phase or phases along the univariant lines form extensive solid solution with a third component lying outside the original compositional join (plane).

A flow sheet summarizing the univariant and invariant characteristics for the quaternary system $A-B-C-D$ is shown in Fig. 9b. Four univariant lines, e_1I, e_2I, e_3I, and e_4I, originating in the limiting ternary systems, meet at the invariant point, which is a quaternary eutectic. At I, solid phases $A + B + C + D$ are in equilibrium with the liquid and temperature, and composition must remain constant until the liquid phase is eliminated. The final mixture at I will be that of the solid phases $A + B + C + D$.† However, in systems with complex phase relations in the limiting ternary systems and planes or joins, many melts may crystallize before reaching the lowest temperature quaternary invariant points.

The presence of intermediate congruently or incongruently melting compounds will complicate, to a certain degree, the path of liquids to the invariant points. Petrological examples of this type are discussed in the subsequent sections.

8 THE SYSTEM DIOPSIDE–FORSTERITE–ALBITE–ANORTHITE

Thus far the phase relations of forsterite or diopside with plagioclase feldspars and silica have been individually and simply presented to demonstrate the equilibrium and nonequilibrium crystallization establishing analogies with natural trends.

8.1 The Limiting Systems

The system Di–Fo–Ab–An is bounded by the ternary systems Di–Fo–An, Di–Ab–An, Di–Fo–Ab, and Fo–Ab–An.

The relations in the system Di–Fo–An have been studied by Osborn and Tait (1952) and are shown in Fig. 10a. It is a pseudoternary due to the presence of the spinel field. If the spinel field is ignored, the system is a simple ternary with a eutectic E* at 1270°C. All melt compositions that lie outside the triangle ABC

---

*Due to diopside being aluminous, the point E is not strictly eutectic, but may be considered so. Natural pyroxenes, as well as those in experimental systems containing aluminous phase, crystallize Al-pyroxenes.

†For systems containing components with complex and extensive solid solutions, the melts may crystallize to only three solid phases at a minimum on the univariant lines (see Schairer, 1942, 1954, for more details) instead of reaching an invariant point with four solid phases. Di–Ne–Ks–Sil is an example of such a case (see Sood et al., 1970).

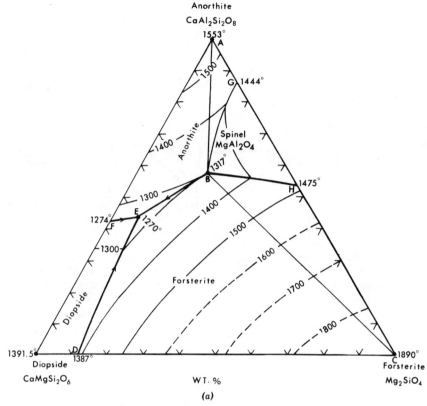

Figure 10 Phase relations in the constituent systems of the diopside–forsterite–albite–anorthite system. *(a)* The system diopside–forsterite–anorthite. (After Osborn and Tait, 1962. With permission of the *American Journal of Science*.)

will begin crystallization with forsterite, diopside, or anorthite. The liquids will then follow the respective phase boundaries *DE, FE,* or *BE,* and the crystallization will terminate at *E,* with temperature and composition remaining constant, giving a final solid mixture Fo + Di + An. The proportion of the final phases will be dependent on the initial bulk composition. Only liquids of composition *E* will separate Di + An + Fo together as the crystallization begins with the temperature and composition remaining constant until the liquid phase is eliminated. Liquids in the triangle *ABC* will leave the system and reenter at *B* after crystallization of spinel. They will then follow the line *BE* to *E* where Di + An + Fo will separate simultaneously.

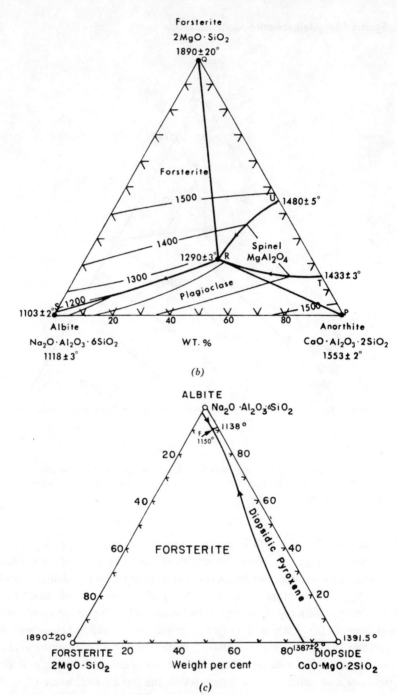

Figure 10 *(b)* The system forsterite–albite–anorthite. (After Schairer and Yoder, 1967. With permission of the Carnegie Institution of Washington.) *(c)* The system diopside–forsterite–albite: the "plane of critical silica undersaturation" (see Fig. 12). (After Schairer and Morimoto, 1958. With permission of the Carnegie Institution of Washington.)

The relations in the system Fo–Ab–An were determined by Schairer and Yoder (1967) and are shown in Fig. 10b. The presence of the field of spinel makes the behavior pseudoternary. A cotectic line RS separates the forsterite and plagioclase fields. The liquids in the triangle PQR reach the cotectic only upon reentry into the system at R after crystallization of spinel. Due to the solid solution only two phases crystallize at some point on the line SR. The crystallization of the compositions outside the triangle PQR (Fig. 10b) will be similar to that in the system Di–Ab–An. The relations in the system Di–Ab–An have been discussed in Section 3.

The system Di–Fo–Ab was studied by Schairer and Morimoto (1959) and the phase relations are reproduced in Fig. 10c. It has an important piercing point F at 1150°C. It is an important plane in the "Basalt Tetrahedron" and will be discussed further in Section 10.

8.2 Phase Relations in the Quaternary System

Let us now examine the quaternary system Di–Fo–Ab–An, which compositionally corresponds to important aspects of *critically silica-undersaturated mafic and ultramafic rocks*. The condensed liquidus phase relations in this system are reproduced in Fig. 11.

Some important features of phase relations in this system are noted below:

1 The presence of the spinel volume indicates that the system is not strictly quaternary.
2 It has prominent phase volumes of plagioclase feldspar solid solution, forsterite, and diopside.
3 The system has two important piercing points, P and Q (Fig. 11), at 1270°C and 1135°C, respectively—a difference of only 135°C.
4 Composition of the piercing points P and Q is as follows:

Piercing Point	Temp. (°C)	Composition (wt. %)			
		Di	Fo	Pl	
P	1270°C	49.0%	7.5%	43.5%	(An-rich)
Q	1135°C	8.0%	1.5%	90.5%	(Ab-rich)

5   The proportion of mineral phases along the four-phase (univariant) curve PQ closely approximates the composition of many olivine basalts.
6   Small differences in the original bulk compositions will change the primary phase (and possibly define different rock associations or types).
7   The temperatures along the univariant curve PQ have close similarities to those of many lavas (see Table XX).

8.3   Crystallization Trends

Depending on the plot of the initial bulk composition, but obviously outside the spinel volume in the diagram (Fig. 11), the liquid will begin crystallization with either the plagioclase feldspar, diopside, or forsterite phases. The melt with continued cooling will approach a divariant surface (three-phase field such as PQR; Fig. 11) and finally reach the univariant line PQ along which Di + Fo + Pl will crystallize. The distance the melt will crystallize along the line PQ, and the final temperature of consolidation will be governed by the initial composition and the location of plagioclase–liquid tie lines similar to the binary and ternary systems containing plagioclase feldspars (see Yoder and Tilley, 1962; Ehler, 1972). However, the tendency with fractionation will be to form an ultramafic to mafic lineage with residual liquids enriched in the sodic component. For melt compositions that crystallize spinel as the first or the second phase (in other words, those that are in, or reach, the spinel volume), the crystallization trend will be somewhat complicated. Such liquids leave the system and the composition moves along appropriate divariant surfaces until it meets the univariant line Fo + Pl + Sp and reenters the tetrahedron at T. Spinel dissolves at T as the liquid moves to P along TP.

The above description has interesting petrological implications with respect to occurrence of spinel in mafic-ultramafic rocks. Spinel is found commonly associated with assemblages such as Ol + En, Ol + En + Di, or An characteristic of mafic-ultramafic sequences. Turner and Verhoogen (1960) suggest the following as a possible reaction to explain the association of spinel:

$$2Mg_2SiO_4 + CaAl_2Si_2O_8 \rightleftharpoons CaMgSi_2O_6 + 2MgSiO_3 + MgAl_2O_4 \quad (1)$$

Due to increase in molar volume to the right the reaction will be favored by high pressure, thus plutonic environments of ultramafic rocks may be suitable.

The presence of spinel-pyroxene coronas around olivine-An-rich plagioclase in troctolites supports such a reaction.

However, the spinel of mafic and ultramafic rocks is chrome-bearing. Irvine (1977) has found that small amount of Cr_2O_3 extends the field of spinel in the system Di–Fo–An–Sil–$MgCr_2O_4$ (Fig. 7c). So reaction (1) may be rewritten as

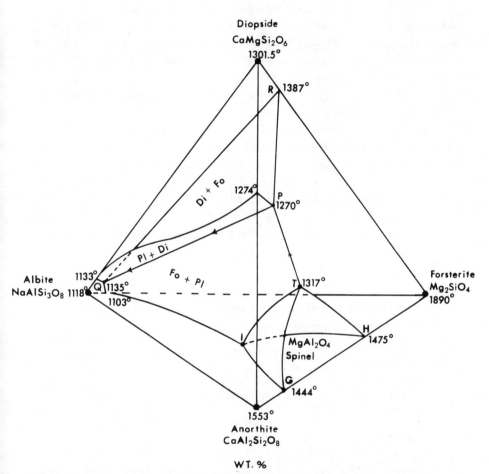

Figure 11 Condensed phase equilibria in the system diopside–forsterite–albite–anorthite representing the critically silica undersaturated basalt compositions. The temperature change along the univariant line PQ is only 135°C. Di + Fo + Pl + Liq are in equilibrium along this line. (After Yoder and Tilley, 1962, with permission of the *Journal of Petrology* and Schairer and Yoder, 1967, with permission of the Carnegie Institution of Washington.)

$$3Mg_2SiO_4 + CaAl_2Si_2O_8 + Cr_2O_3 \rightleftharpoons CaMgSi_2O_6$$
$$+ 3MgSiO_3 + 2Mg(AlCr)O_4 \quad (2)$$

Obviously if iron is present then the $Mg^{2+} \rightleftharpoons Fe^{2+}$ substitution could take place under the right oxygen pressure conditions.

Irvine observed a reasonable agreement in the Cr/Cr + Al ratios of the spinel phase on the crystallization paths in the synthetic system Di–Fo–An–Sil –MgCr$_2$O$_4$ (Fig 7d) and those from tholeiites and layered intrusions (Sigurdsson and Schilling, 1976; Evans and Wright, 1972; Irvine, 1967; Jackson, 1963, 1969; Worst, 1960; Cameron, 1969, 1971). A summary of Cr/Cr + Al ratios for various petrologic types and those of fractionation paths in the synthetic system is given in Table IV.

The data indicate a somewhat direct relationship between the Cr/Cr + Al ratios, degree of differentiation, and bulk chemistry of the rock. The spinels from ultramafic rocks tend to be more Cr-rich than those in mafic rocks.

From the consideration of phase relations in the system Di–Fo–Ab–An an important conclusion appears to be in order.

For compositions outside the spinel volume, irrespective of the initial composition or the separation of a particular primary phase, the other phases begin

Table IV Cr/Cr + Al Ratios of Chromites in Mafic-Ultramafic Rocks

Rock Type	100 Cr/Cr + Al Ratio	Crystallization Path Numbers[a] (Fig 7d)
Abyssal tholeiites	37–54 (low)	1
Kilauean tholeiites, Hawaii	56–74 (moderate)	2
Muskox intrusion, N.W. Territories, Canada	70–77 (moderate–high)	3
Stillwater complex, Montana, Bushveld complex, S. Africa, Great Dyke, Rhodesia	78–82 (high)	4

[a]From Irvine (1977).

to crystallize only with a small drop in temperature. In other words, shortly after the crystallization of a primary phase, liquids approach the four-phase curve PQ.

This implies that basaltic (or magma) compositions must lie close to the line PQ (Fig. 11), or a four-phase curve. As fractional melting follows equilibrium curves, it is suggestive that basalts themselves may be products of fractional melting (or crystallization). Melting experiments on basalts (Yoder and Tilley, 1962) and phonolites/syenites (Sood and Edgar, 1970; Piotrowski and Edgar, 1970) show crystallization of four phases for many diverse compositions at about the same temperature or over very short temperature intervals. This is in accord with the observation that liquids on fractionation or compositional lines must reach a four-phase curve at about the same temperature (see Osborn and Schairer, 1941; Ehler, 1972; Ricci, 1966; Zavaritskii and Sobolev, 1964). Therefore it indicates that basalt types (or magma types) may be related by subtraction of a phase or phases of fixed (variable) compositions normally crystallizing from basalts (or such magmas). In other words, different basalts (parental basalts) may be related by fundamental or basic differences in composition of the source material (Yoder and Tilley, 1962; Schairer, 1967; Yoder, 1976).

9 THE SYSTEM DIOPSIDE–FORSTERITE–NEPHELINE–SILICA: THE SIMPLE BASALT TETRAHEDRON

The foregoing discussion permits now an analysis of the collective behavior of mafic and salic minerals to establish liquid lines of descent characterizing silica-undersaturation and -oversaturation lineages of rock series and their spatial separation in nature.

The experimental work of Yoder and Tilley (1962) on natural and synthetic basalt compositions was a landmark contribution in this direction. This opened new vistas for petrological research and provided explanation for many aspects of basalt crystallization and genesis. In particular the study emphasized a continuum that exists in various basalt compositions and the dependency of a basalt composition on the depth and degree of partial melting in the mantle, for example, alkali-basalt magmas probably form at a greater depth than tholeiitic type of magmas.

In the preceding description of various systems, nepheline, an important constituent of many alkali basalts, was not represented as a component. The system Di–Fo–Ne–Sil (Fig. 12), the iron-free "Simple Basalt System" of Yoder and Tilley (1962), which includes nepheline, not only accommodates the compositional variants of basalts (see Fig. 13) but also represents the principal phases of basaltic rocks.

9.1 Features of the Tetrahedron Di–Fo–Ne–Sil

Though certain compositional limitations exist, still the Basalt Tetrahedron (Fig. 12) has great relevance to mafic rocks. It may be appropriate to summarize from the study of Yoder and Tilley (1962) the following features of the tetrahedron:

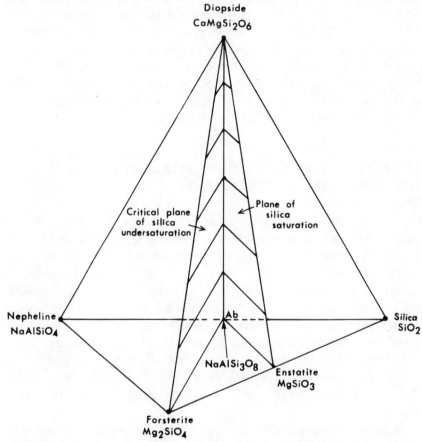

Figure 12 The iron-free simple Basalt Tetrahedron diopside–forsterite–nepheline–silica conforming to normative compositions of basalts. (After Yoder and Tilley, 1962. With permission of the *Journal of Petrology*.)

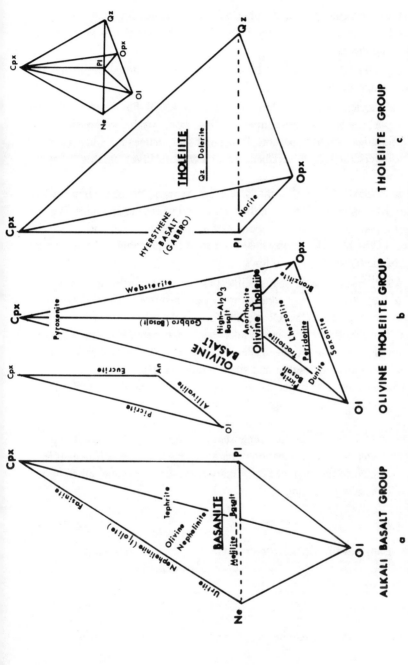

Figure 13 The Basalt Tetrahedron showing rocks which can be normatively represented by the system (see Table V). Those compositions which lie in the face are parallel to the bases, whereas those underlined plot within the tetrahedron. (*a*) The critically silica undersaturated portion Cpx–Ne–Ol–Pl representing alkali basaltic compositions (*b*) The portion Cpx–Ol–Opx–Pl representing olivine–tholeiite compositions. (*c*) The silica-oversaturated portion Cpx–Pl–Opx–Qz representing tholeiitic–quartz tholeiitic compositions. (After Yoder and Tilley, 1962. With permission of the *Journal of Petrology*.)

43

1. It has two important planes: the plane Di–Fo–Ab called the "Plane of Critical Silica Undersaturation," and the plane Di–Ab–En referred to as the "Plane of Silica Saturation."
2. These planes (Di–Fo–Ab and Di–Ab–En) divide the tetrahedron into two major compositional units.
   a) The critically silica-undersaturated portion, Di–Fo–Ne–Ab, encloses nepheline- and olivine-normative, silica-poor basaltic compositions.
   b) The silica-saturated portion, Di–En–Ab–Sil, defines the silica-saturated to oversaturated (hypersthene or quartz normative) basaltic compositions.
3. The tetrahedron Di–Fo–Ne–Sil is relevant to many diverse compositions. Figure 13 is an enlarged view of the tetrahedron where general mineral group names replace the specific components for broader representation. Included in Fig. 13 are the rock names whose normative compositions lie in the various parts of the tetrahedron as follows.
   a) Alkali basalts, basanites, melilite basalts, nephelinites, urtites-ijolites, and tephrites defining alkali affinities are restricted to the silica-undersaturated portion Cpx–Ol–Ne–Pl (Fig. 13a).
   b) Olivine tholeiites, olivine basalts, and fractionates from such compositions are limited to the space Cpx–Ol–Opx–Pl bounded by the planes of silica undersaturation and silica saturation (Fig. 13b).
   c) Quartz tholeiites, hypersthene basalts, and their derivatives plot in the silica-oversaturated volume Cpx–Opx–Pl–Sil (Fig. 13c).

Yoder and Tilley (1962) thus suggested a compositional continuity among basalts (see Table V for classification of basalts), whereas field petrologists have generally maintained the view of two primary basalts, for example, alkali basalts and tholeiite basalts producing

alkali basalt–hawaiite–mugearite–trachyte ⟨ phonolite / pantellerite *(the alkali trend)*

and

tholeiite–andesite–dacite–rhyolite *(the silica-enrichment trend)*

Table V Normative Classification of Basalts[a]

1. Tholeiitic types
   - Quartz tholeiites (silica-oversaturated) — Hypersthene and quartz-normative
   - Tholeiites (silica-saturated) — Hypersthene-normative
   - Olivine tholeiites (silica-undersaturated) — Olivine and hypersthene-normative
2. Basaltic types
   - Olivine basalts (silica-undersaturated) — Olivine-normative
   - Alkali basalts (silica-undersaturated) — Olivine and nepheline-normative
   - basanites (silica-undersaturated) — Olivine and nepheline-normative
   - Olivine nephelinites (highly silica-undersaturated) — Olivine and nepheline-normative (essentially feldspar-free)
3. High alumina basalts (silica-undersaturated) — Olivine and/or nepheline-normative (contain up to 18% Al_2O_3)

[a]Modified from Yoder and Tilley (1962).

9.2 Phase Relations in the System Di–Fo–Ne–Sil

It will be useful to describe the phase relations in the constituent ternary systems Di–Ne–Sil, Di–Fo–Ne, Fo–Ne–Sil, and Di–Fo–Sil before considering the relations in the quaternary system.*

---

*In actual fact the behavior in this system is not strictly quaternary, as the oxides behave, in part, as independent components. The relations can be adequately considered in the system $CaO-MgO-Na_2O-Al_2O_3-SiO_2$. However, as the extent of independent behavior of oxides is somewhat limited, so the system may be treated as a quaternary and terms related to quaternary representation are used. A similar basis is used for the other quaternary systems discussed.

The system Fo–Ne–Sil was studied by Schairer and Yoder (1961), and its phase relations are shown in Fig. 14a. The join Fo–Ab is an equilibrium thermal divide and separates the system into silica-poor (Fo–Ne–Ab) and silica-rich (Fo–Ab–Sil) portions. This implies that compositions on either side of the join must crystallize in the respective compatibility (compositional) triangles and cannot cross the "thermal divide." Therefore a very slight variation in composition from that of the Fo–Ab join will markedly change the crystallization paths of the liquids. Fa–Ne–Sil (Bowen and Schairer, 1935), Di–Ne–Sil (Schairer and Yoder, 1960a), An–Ne–Sil (Schairer, 1954), and other systems (see Fig. 8) containing forsterite or nepheline or leucite with silica show similar thermal divides.

The system Di–Ne–Sil (Fig. 14b) has been studied by Schairer and Yoder (1960a), where the join Di–Ab divides it into silica-rich (Di–Ab–Sil) and silica-poor (Di–Ne–Ab) portions.

Di–Ab–Sil contains an invariant point I at 1073°C, whereas Di–Ne–Ab has two invariant points, G and H. Of particular interest is the Fo + Liq reaction point G producing diopside (cf. Ol + Mel reaction in the system Di–Ne). Schairer and Yoder (1960a, p. 281) suggest that such a reaction might be important in the genesis of basanites, melilite basalts, and olivine nephelinites, although Wilkinson (1956) believes these reactions to be absent in alkaline basalts.

The system Di–Fo–Ne (Fig. 14c) was studied by Schairer and Yoder (1960b). There is a large phase area of forsterite penetrating the Di–Ne join. The invariant point A is not important for the present discussion due to the presence of spinel. The point B does not have an assemblage analogous to natural occurrences. There is a minimum C on the Di–Ne line at 1260°C. The system is pseudoternary.

The system Di–Fo–Sil was first studied by Bowen (1914). Kushiro and Schairer (1963) have obtained more detailed data on certain portions of the system. The updated phase relations are shown in Fig. 14d. It has two peritectics, point A at 1386°C and point B at 1374°C.

Besides the limiting ternary systems just described, the data on the planes Di–Fo–Ab and Di–En–Ab are also available.

The system Di–Fo–Ab, the plane of critical silica undersaturation, was studied by Schairer and Morimoto (1959) and is reproduced as Fig. 10c. The system is a pseudoternary with an important piercing point F at 1150°C. For many compositions containing less than 2.0 wt. % forsterite, it is still the first phase to separate.

Di–Fo–Ab is an equilibrium "thermal divide" in the system Di–Fo–Ne–Sil

The System Diopside-Forsterite-Nepheline-Silica

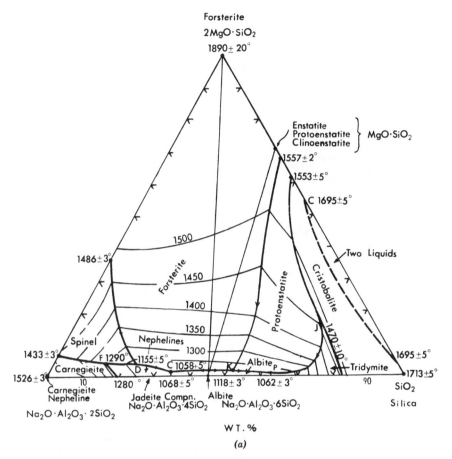

Figure 14 Phase equilibrium relations in the various constituent systems and planes in the Basalt Tetrahedron. Note large fields of the mafic components. *(a)* The system forsterite–nepheline–silica: the base of the tetrahedron. The join Fo–Ab is an equilibrium thermal divide. (After Schairer and Yoder, 1961. With permission of the Carnegie Institution of Washington.)

at 1 atm pressure. The melt compositions on the nepheline side of this plane will produce nepheline-bearing assemblages, whereas those on the silica-rich side will form quartz and/or hypersthene-bearing assemblages.*

---

*In other words, under low pressure equilibrium conditions, one basalt magma cannot give rise to both alkali basalt (silica-undersaturated) and tholeiite (silica-oversaturated) trends at the same time (Yoder and Tilley, 1962).

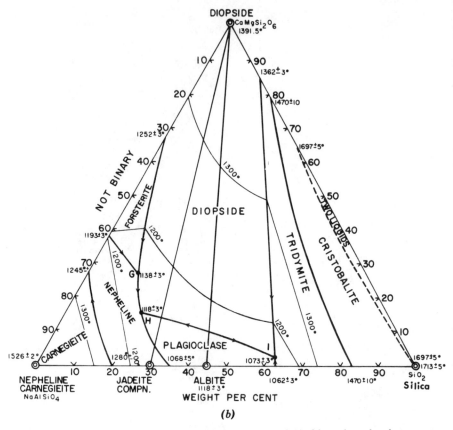

Figure 14 *(b)* The system diopside–nepheline–silica. Note the field of forsterite makes the system a pseudoternary. The join Di–Ab separates the system into silica-undersaturated and -oversaturated portions. (After Schairer and Yoder, 1960a. With permission of the *American Journal of Science*.)

The effect of the thermal divide, however, could be weakened or shifted to a certain degree by factors such as Fo + Liq or other reactions, the plagioclase effect forming Al-pyroxenes, oxidation reactions, the pressure effect, and others. (For further treatment see Yoder, 1976; Yoder and Tilley, 1962; Schairer and Yoder, 1964.)

The phase relations in the system Di–En–Ab and the plane of silica saturation are shown in Fig. 14*e* (Schairer and Morimoto, 1958). The piercing point *K* has similar composition as for Di–Fo–Ab (see Fig. 10*c*), but the temperature is somewhat lower (1138°C), which further reinforces Di–Fo–Ab as a thermal divide. This implies that the liquids on the nepheline side of the Di–Fo–Ab plane have, through fractionation, the capacity to produce silica-rich derivatives. The reverse, however, does not seem likely from the phase relations in the system Di–Fo–Ne–Sil.

9.3 Flow Sheet with Univariant and Invariant Relations

The univariant and invariant relations in the system Di–Fo–Ne–Sil can be represented in the form of a flow sheet shown in Fig. 15. The data on invariant and

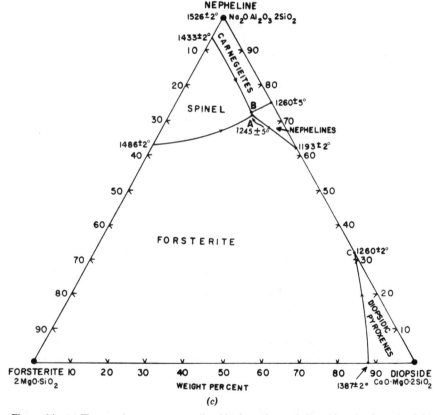

Figure 14 (c) The pseudoternary system diopside–forsterite–nepheline. Note the forsterite field penetrates the Di–Ne join. (After Schairer and Yoder, 1960b. With permission of the Carnegie Institution of Washington.)

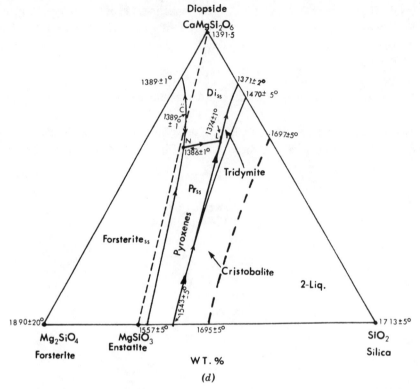

Figure 14 *(d)* The system diopside–forsterite–silica (cf. Fig. 6a). (After Kushiro and Schairer, 1963. With permission of the Carnegie Institution of Washington.)

piercing points for the origin of the univariant lines in the limiting systems and the joins of the system Di–Fo–Ne–Sil are given in Table VI. The arrows point in the direction of falling temperature. The divergent arrows indicate a thermal divide. Solid phases listed are those coexisting with the liquid along the univariant lines and at the invariant points. The systems are shown in parentheses for the univariant lines. The letters and temperatures refer to those of the invariant and piercing points. Such a representation is useful in charting the liquid lines of descent.

In the system Di–Fo–Ne–Sil there are three quaternary invariant points: A, B, and C (Fig. 15). Points A and C, lying in the portions Di–Fo–Ab–En and Di–Fo–Ne–Ab (Fig. 12), are both peritectics and involve forsterite reaction relation. The third point B is a eutectic and lies in the portion Di–En–Ab–Sil.

9.4 Liquid Trends

Let us now consider some possible crystallization trends for compositions in this system with the help of the flow sheet.

The liquid composition (or assemblage) at point *A* (Fig. 15) represents a simplified olivine tholeiite. Upon dissolution of olivine at *A*, it passes through a hypersthene basalt along the line *AB* to a quartz tholeiite at point *B*. This illustrates that a magma of olivine tholeiite composition could give rise, upon fractionation, to the trend *olivine tholeiite–hypersthene basalt–quartz tholeiite*.

Liquid compositions on the plane Di–Fo–Ab (Fig. 12), which correspond to simple olivine basalts, will go to points *A* or *C* (Fig. 15). They will reach point *C* only if they are critically silica-undersaturated, otherwise they will move

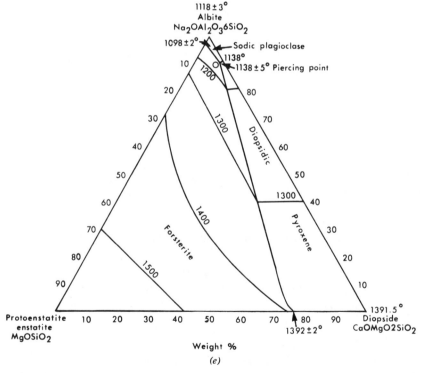

Figure 14 *(e)* The system diopside–enstatite–albite: the plane of silica saturation. (After Schairer and Morimoto, 1958. With permission of the Carnegie Institution of Washington.)

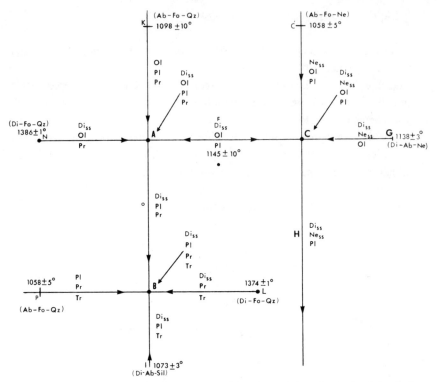

Figure 15 Flow sheet showing the univariant and invariant relations for the basalt tetrahedron diopside–forsterite–nepheline–silica. Arrows indicate direction of falling temperatures. Letters represent the piercing point in the constituent systems and planes shown in the parentheses. The three solid phases in equilibrium with liquid are shown on the univariant lines. Point B is a quaternary eutectic in the silica-rich portion of the system. (Adapted from Schairer and Yoder, 1964. With permission of the Carnegie Institution of Washington.)

to point A. In both cases they will undergo forsterite (olivine) dissolution at A and C.

Those liquids, which upon crystallization pass through point C, will result in the *olivine basalt–nepheline basanite–nepheline tephrite* trend.

Liquids reaching point A will dissolve forsterite at A and terminate crystallization at B, forming *olivine basalt–olivine tholeiite–hypersthene basalt–quartz tholeiite* association.

Liquids near the Di–Fo–Ne plane (Fig. 12) may also undergo olivine dissolution at C (Fig. 15) and form *olivine nephelinite–nepheline*

Table VI Piercing Points of the Univariant Lines in the Various Portions of the System Di–Fo–Ne–Sil: Simple Basalt Tetrahedron[a]

System	Relevant Subtrahedron (in Fig. 12)	Important Piercing Point(s)	Solid Phases[b]	Temp. (°C)	Quaternary Invariant Points (Four Solid Phases,[b] Fig. 15)
Di–Fo–Ne (Fig. 14c)[d]	Critically silica-undersaturated portion (Di–Fo–Ne–Ab)	A	Fo + Ne + Sp	1245 ± 5	
		B	Ne + Sp + Cg	1243 ± 5	
Fo–Ne–Ab (Fig. 14a)		C'	Fo + Ne + Ab	1058 ± 5	
		D	Fo + Ne + Sp	1155 ± 5	C (olivine peritectic) Di + Ol + Ne + Pl
Di–Ne–Ab (Fig. 14b)	Silica undersaturated To saturated portion (Di–Fo–En–Ab)	H	Di + Ne + Pl	1118 ± 3	
Di–Fo–Ab (Fig. 10c)		G	Di + Fo + Ne	1138 ± 3	
Di–Fo–En (Fig. 14d)		F	Di + Fo + Pl	1150 ± 5	
		C''	Di + Fo	1389 ± 1	
Fo–En–Ab (Fig. 14a)		—	—	—	A (olivine peritectic) Di + Ol + Pl + Pr
Di–En–Ab (Fig. 14e)	Silica-saturated to -oversaturated portion (Di–En–Ab–Sil)	O	Di + Fo + Pl	1138 ± 5	
En–Ab–Sil (Fig. 14a)		K	Fo + Pr + Ab	1098 ± 5	
Di–Ab–Sil (Fig. 14b)		P	Pr + Ab + Trid	1058 ± 5	
		I	Di + Pl + Trid	1073 ± 3	B (eutectic) Di + Pl + Pr + Tr
Di–En–Sil (Fig. 14e)		N	Fo + Pr + Di	1386 ± 1	
		L	Di + Pr + Trid	1374 ± 1	

[a]Check text for references. [b]Solid phases co-exist with liquid. [c]Essentially represent ol (olivine). [d]Figure numbers are those in the text. For explanation of abbreviations see List of Abbreviations.

basanite–nepheline tephrite trend. Such liquid lines of descent are common in alkali provinces of African Rift System (Wright, 1963, 1971; Saggerson and Williams, 1964), Napak area of Uganda (King, 1965), and Bohemian Mittelgebirge (Knorr, 1932; Scheuman, 1913). Turner and Verhoogen (1960) and Hyndman (1972) also mention many other examples of similar alkali rock associations.

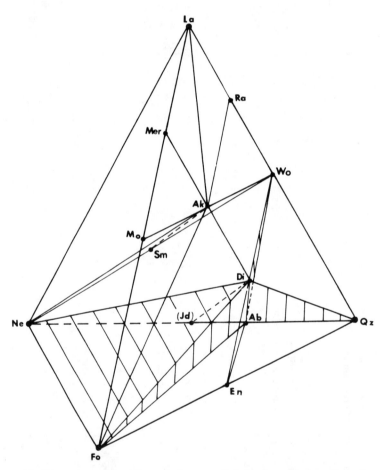

Figure 16 Schematic representation of the expanded basalt tetrahedron forsterite–nepheline–larnite–silica, where the simple Basalt Tetrahedron forms the base. Such expansion includes the melilite-bearing basalts. Ruled planes are "almost thermal divides." (After Schairer and Yoder, 1964. With permission of the Carnegie Institution of Washington.)

10 THE SYSTEM FORSTERITE–NEPHELINE–LARNITE–SILICA: THE EXPANDED BASALT TETREHEDRON

The above discussion of phase relations in the system Di–Fo–Ne–Sil clearly demonstrates its usefulness in explaining the crystallization behavior of basaltic magmas. The absence of melilite as a component, however, limits its broad applicability. Schairer and Yoder (1964) constituted the system Fo–Ne–La–Sil, which is an expanded version of the basalt tetrahedron. This includes, in addition to melilites, most members of the solid solution mineral series. It also permits consideration of the plagioclase-melilite incompatibility found in rocks.

The expanded tetrahedron with petrologically important joins is shown in Fig. 16. In addition to various joins and the limiting system of the simple Basalt Tetrahedron discussed before, Schairer and Yoder (1964) also provided data on six joins (Di–Ne–Ak, Ab–Wo–Ak, Ne–Ak–Wo, Ne–Ak–Ab, Ab–Ak–Di, and Ne–Wo–Di; see Fig. 17*a*–*f*) in the Fo–Ne–La–Sil system, which have implications to the melilite-bearing rocks. The relevant phase equilibrium data on the piercing points in the joins and expanded tetrahedron are listed in Table VII, excluding those listed in Table VI.

10.1 Flow Sheet with Univariant and Invariant Relations

On the basis of extensive data collected on many subsystems (see Tables VI and VII) it was possible for Schairer and Yoder (1964) and Schairer (1967) to present the univariant and invariant relations in this system in the form of a flow sheet. The detailed flow sheet is reproduced here as Fig. 18*a*.

The solid phases coexisting with the liquid are listed along the univariant lines. The arrows indicate the direction of falling temperatures. The univariant and invariant relations are important as they dictate the physicochemical behavior of melts (magmas) and explain possible liquid trends.

Some important aspects of the flow sheet can be summarized with the help of Fig. 18*a,b*.

1 In Fig. 18*a* the portion to the left of the line *PQ* schematically represents the univariant and invariant relations in the silica-saturated and -oversaturated portion of the system Fo–Ne–La–Sil. The portion to the right of the line *PQ* is relevant to the relations in the silica-undersaturated portion. This silica undersaturation increases to the right of the diagram.

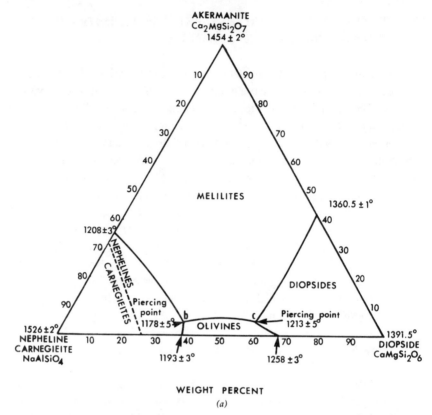

Figure 17 Phase equilibrium relations in the six joins (17a–f) in the petrologically important portion of the expanded Basalt Tetrahedron. (After Schairer and Yoder, 1964. With permission of the Carnegie Institution of Washington.) *(a)* The join diopside–nepheline–akermanite. There are two piercing points. Point b, at 1178 ± 5°C, has Ol + Mel + Ne + Liq, whereas point c, at 1213 ± 5°C, has Ol + Di + Mel + Liq. (After Schairer, 1967. With permission of John Wiley & Sons, Inc.)

2. In the silica-saturated to -oversaturated portion of the tetrahedron, the invariant points B and C (Fig. 18a) are quaternary eutectics, whereas point A is a reaction point for forsterite.

3. In the silica-undersaturated portion of the system the invariant point E (Fig. 18a) is the only quaternary eutectic. The invariant points F and G are reaction points for melilite and D for forsterite. The assemblages along the line FG are feldspar-free.

The System Forsterite-Nepheline-Larnite-Silica

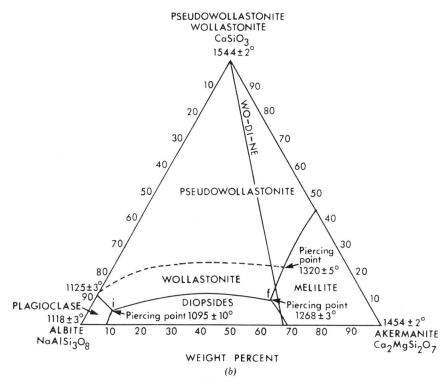

Figure 17 *(b)* The join albite–wollastonite–akermanite. Note the piercing points *f*, Mel + Wo + Di + Liq at 1268 ± 3°C, and *i*, Di + Wo + Pl + Liq at 1095 ± 10°C showing separation of plagioclase and melilite fields. (After Schairer, 1967. With permission of John Wiley & Sons, Inc.)

4. The four planes Di–Fo–Ab, Di–Fo–Ne, Di–Ab–Sil, and Di–Ab–Wo (Fig. 18*a,b*) are the thermal divides in the system. The planes Di–Fo–Ab and Di–Fo–Ne intersecting the univariant lines *AD* and *DF*, respectively, are, however, petrologically important.

5. The temperature maxima on the planes Di–Fo–Ab and Di–Ab–Wo (Fig. 16), represented by the divergent arrows on the lines *AD* and *CE* (Fig. 18*a*), separate the invariant points *A*, *B*, and *C* in the silica-saturated (nepheline-free) portion from the invariant points *D*, *E*, *F*, and *G* in the critically silica-undersaturated (nepheline-bearing) portion of the system. They essentially differentiate the silica-rich and silica-poor liquids.

6. The temperature maximum on the univariant line *DF* (Fig. 18*a*) in the silica-undersaturated portion separates the invariant points *D* and *E* from the points

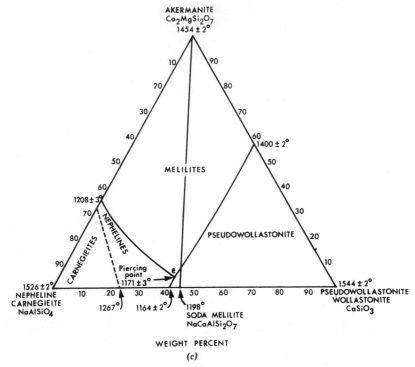

Figure 17 *(c)* The join nepheline–wollastonite–akermanite. Note the piercing point *e*, at 1171 ± 3°C, where Wo + Ne + Mel + Liq are in equilibrium. (After Schairer, 1967. With permission of John Wiley & Sons, Inc.)

F and *G*. In other words, it differentiates the plagioclase-bearing from the melilite-bearing compositions; this suggests a reason for plagioclase-melilite incompatibility in nature.

7 The temperature high on the line *CE* separates the quartz-bearing from the nepheline-bearing compositions at the eutectics *C* and *E*.

8 There are two petrologically important eutectics, *B* and *E*. The eutectic *B* is the goal for all liquids in the silica-rich portion, whereas *E* is the major goal for liquids in the silica-poor portion. The reaching of *B* or *E* will be facilitated only by a very high degree of fractionation.

10.2 Liquid Trends

It may be appropriate at this point to summarize the probable liquid lines of descent for an alkali basalt magma in the light of relations in the expanded Basalt

The System Forsterite-Nepheline-Larnite-Silica

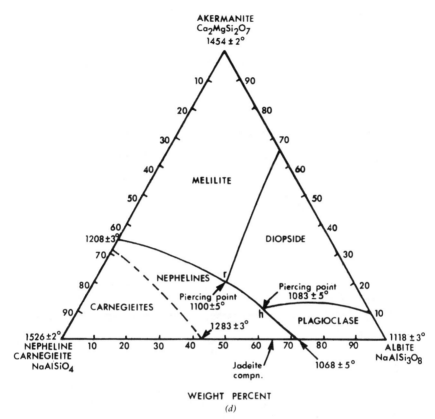

Figure 17 *(d)* The join nepheline–akermanite–albite. There are two important piercing points. Point *r*, at 1100 ± 5°C, has Di + Mel + Ne + Liq, and point *h*, at 1083 ± 5°C, contains Di + Ne + Pl + Liq in equilibrium (cf. system Di–Ne–Sil). Note antipathy of melilite and plagioclase. (After Schairer, 1967. With permission of John Wiley & Sons, Inc.)

Tetrahedron, which is considered by field geologists to be the parent magma type for the alkali succession. Alkali basalt is principally composed of clinopyroxene, plagioclase, and olivine, with small amounts of nepheline.

The plane Di–Fo–Ab is an important compositional plane and the likely site, according to Tilley and Yoder (1964), where alkali basalt magmas can be generated. Based on the extent of silica undersaturation and sodic content of the melt,* such liquids may pass through the invariant point *D* and, with extensive fractionation, terminate crystallization at *E* (Fig. 18*b*). This will define the trend:

---

*The melts approaching An_{50} (or more) plagioclase composition are not likely to move toward the basanite invariant point to give nepheline-bearing rocks.

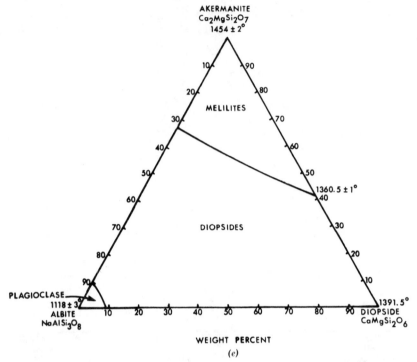

Figure 17 *(e)* The join albite–akermanite–diopside. Note that the diopside field intersects the albite–akermanite join. Thus there is no piercing point but cotectics separating the fields of diopside–akermanite and diopside–albite. This shows incompatibility of melilite and plagioclase. (After Schairer, 1967. With permission of John Wiley & Sons, Inc.)

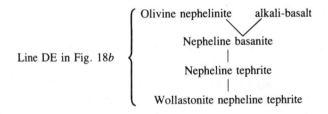

It is probable that wollastonite will be present as sphene or perovskite in a natural lineage. The liquid trend to tholeiite and quartz tholeiite is already discussed in the previous section.

The eutectic E (Fig. 18a,b) can also be reached by another less important course of fractionation recorded by a less common group of melilite-bearing rocks. The liquids of the composition corresponding to an olivine nephelinite and melilite basalt could be generated on or near the planes Di–Fo–Ne and Di–Fo–Ak, respectively. These liquids, based on the degree of silica-undersaturation, move to point F, which has an assemblage representative of an olivine melilite nephelinite. At F the temperature and composition must remain constant as olivine dissolves. With the elimination of the last of the olivine, the liquid moves toward G. The assemblage along the line FG (Fig. 18a,b) defines a melilite nephelinite. At G wollastonite comes in, and an equivalent rock could be called a wollastonite melilite nephelinite, which is extremely rare. At this point in the fractionation scheme melilite is eliminated by reaction at G, and the

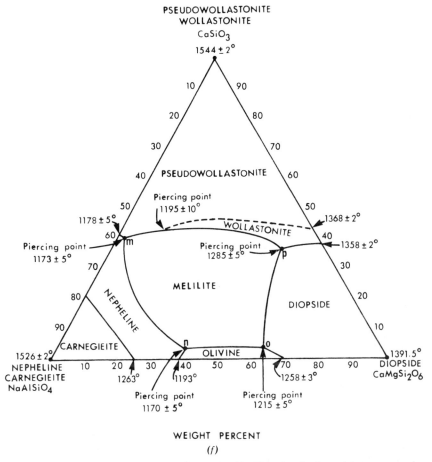

Figure 17 (f) The join nepheline–wollastonite–diopside. Note the piercing points m, n, o, and p are penetrations of points in the joins Di–Ne–Ak, Ab–Wo–Ak, and Ne–Ak–Wo shown in Fig. 17a,b,c. (After Schairer, 1967. With permission of John Wiley & Sons, Inc.)

Table VII Piercing Points in the Portion of the System Di–Fo–La–Sil Expanded Basalt Tetrahedron[a]

System	Relevant Subtetrahedron (SiO$_2$-Saturation Increases Top to Bottom)	Important Piercing Point(s) (Fig 18a)	Three-Solid Phases	Temp. (°C)	Quaternary Invariant Point (Fig. 18a, b) (Four Solid Phases)
Di–Fo–Ak		a	Ol + Mel + Di		
Fo–Ne–Ak	Di–Ne–Ak–Fo	b'	Al + Ne + Mel		F (Olivine peritectic) Di + Mel + Ne + Ol
Di–Fo–Ne		b	—	—	
Di–Ne–Ak (Fig. 17a)[b]		c	Ol + Mel + Ne	1178 ± 5	
			Di + Ol + Mel	1213 ± 5	
		g	Di + Ne + Ol	1138 ± 3	
Ne–Wo–Ak (Fig. 17c)	Di–Ne–Ak–Wo	e	Mel + Ne + Wo	1171 ± 3	G (Melilite peritectic) Di + Mel + Ne + Wo
Di–Wo–Ak (Fig. 17b)		f	Di + Mel + Wo	1350 ± 5	
Di–Ne–Wo (Fig. 17f)		m	Ne + Wo – Mel	1173 ± 5	
		n	Ne + Ol + Mel	1170 ± 5	
		o	Di + Mel + Ol	1215 ± 5	
		p	Di + Mel + Wo	1285 ± 5	
					E (Eutectic) Di + Ne + Pl + Wo
Ne–Ab–Wo	Di–Ne–Ab–Wo	q	Ne + Pl + Wo	1080 ± 5	
		g	Di + Fo + Ne	1138 ± 3	
Di–Ne–Ab (Fig. 14b)		H	Di + Ne + Pl	1118 ± 3	
Di–Wo–Ab (Fig. 17b)		i	Di + Pl + Wo	1095 ± 10	
Di–Wo–Sil		j	Di + Tr + Wo	1320 ± 5	C (Eutectic) Di + Pl + Wo + Trid
Wo–Ab–Sil	Di–Ab–Wo–Sil	k	Pl + Tr + Wo		
Di–Ab–Sil (Fig. 14b)		l	Di + Pl + Trid	1073 ± 3	
Ak–Ne–Ab (Fig. 17d)		r	Di + Mel + Ne	1100 ± 5	

[a]Based on the data from Schairer and Yoder (1960, 1961, 1964), Ferguson and Merwin (1919), and Yoshiki and Yoshida (1952). See text for appropriate references.
For explanation of abbreviations, see List of Abbreviations.

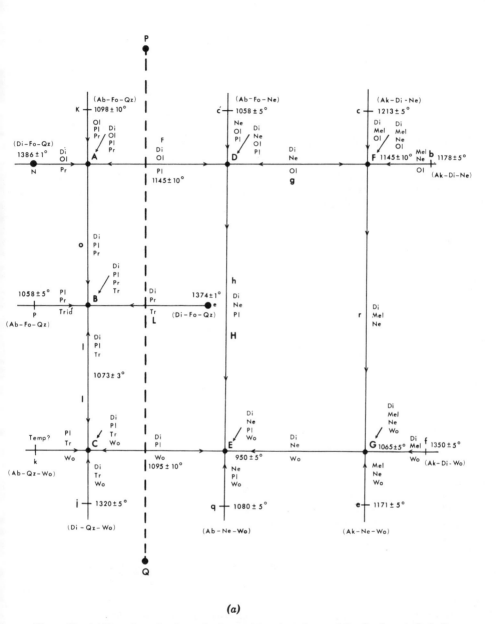

(a)

Figure 18 (a) Flow sheet showing univariant and invariant characteristics for the petrologically important portion of the expanded basalt tetrahedron. Arrows indicate direction of falling temperatures. The systems or joins, the piercing points and their temperatures are listed at the beginning of the univariant lines. The three solid phases in equilibrium with liquid are listed on the univariant lines. Note the three quaternary eutectics. The line PQ schematically divides silica oversaturated and undersaturated portions on the flow sheet. The various possible liquid trends are discernible from the figure. (After Schairer and Yoder, 1964. With permission of the Carnegie Institution of Washington.)

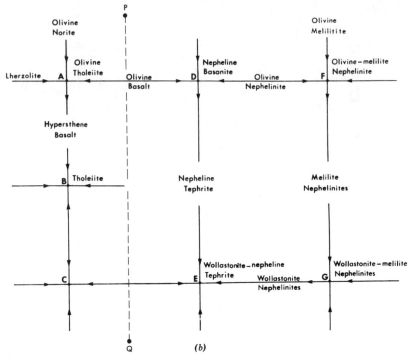

Figure 18 *(b)* Flow sheet with rock names corresponding to assemblages in Fig. 18*a*. (After Schairer and Yoder, 1964. With permission of the Carnegie Institution of Washington.)

liquid moves toward E. The mineral phases along GE are analogous to wollastonite nephelinites. At E the liquid phase disappears with the crystallization of Di + Ne + Wo + Pl. The total fractionation trend possible from magmas near Di–Fo–Ne plane is shown below:

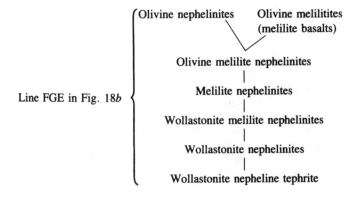

This is a unique example of extensive fractionation. The above discussion leads to an important observation that the liquid that initially gave rise to melilite can upon reaction and fractionation yield a plagioclase (Schairer, 1967).

10.3 Plagioclase–Melilite Incompatibility

The above discussion clearly demonstrates that silica-poor liquids in the system Fo–Ne–La–Sil lead to either plagioclase- or melilite-bearing assemblages over a great part of their fractionation sequence (lines FG or DE, Fig. 18a,b). Therefore it is seen that akermanitic-melilite and plagioclase are incompatible in synthetic silicate systems (see Fig. 17) as well as in rocks. It is possible that this incompatibility is a function of the compositional characteristics of the Di–Fo–Ne plane (or parental magma). Because of this observation, olivine nephelinites are of tremendous importance in ascertaining whether melilite or plagioclase successions of basalts have common parent at depth or their incompatibility disappears with depth. These data suggest that the plagioclase and melilite successions in lavas must form independently (Schairer, 1967).

In recent years there has been an extensive systematic study of the chemistry and petrography of melilite-bearing rocks (see El Gorsey and Yoder, 1973; Yoder, 1973; Velde and Yoder, 1976, 1977; Yoder and Velde, 1976). These data stress that melilite-bearing rocks fall into two categories: (1) the alkali-rich group, which corresponds to a point at 40% nepheline in the subsystem Di–Ne–Ab–Ak, and (2) the magnesian-rich group, to a point at 25% nepheline in the subsystem Ak–Di–Ne–Fo of the expanded Basalt Tetrahedron. Melilite-bearing rocks are poor in Al_2O_3 and SiO_2 and have overlapping CaO contents when compared to nephelinites, alkali basalts, and so on, but still are in a similar compositional domain (Velde and Yoder, 1977), which is important in establishing their genetic lineage to the other basic rocks. On the basis of chemical and selected experimental data it appears that the alkali content, either primary or secondary due to the addition of alkalis to the initial magma, is the principal factor controlling the crystallization of melilites. Therefore, the melilite-plagioclase incompatibility is inherent in the original magma composition, which is not fully definable as yet owing to the presence of melilite in both sodic and potassic mafic rocks.

11 SELECTED MAFIC SILICATE SYSTEMS WITH IRON OXIDES

Thus far the various silicate systems related to basalts have not included iron oxides or iron silicates as components. However, iron is an important constituent of basaltic rocks. The oxidation state of iron in basaltic minerals is related to the prevailing partial oxygen pressure (P_{O_2}) conditions in the magma chamber and is critical to the crystallization of certain minerals (Ol, Px, Sp) and possibly the crystallization trends in general.

At this point it may be useful to discuss phase relations in selected systems containing iron oxides or silicates to more closely approximate natural compositions. Due to the difficulty of controlling the valence state of iron, only limited data on iron-bearing systems are available. The studies of Bowen (1915b); Bowen et al. (1933); Roedder (1951); Bowen and Schairer (1935); Speidel and Osborn (1969); Muan (1958), Muan and Osborn (1956); Osborn and Muan (1960); Phillips and Muan (1959); Roeder (1960); Presnell (1966); Roeder and Osborn (1966); Osborn (1957, 1959, 1962, 1976, 1978); Macleans (1969); Osborn and Arculus (1975); and Osborn and Watson (1977) are the major sources of information on iron-bearing oxide and silicate systems related to basalts.

11.1 The System Forsterite–Wollastonite–Iron Oxide–Silica

Osborn (1962), using his and other published data on the limiting systems, presented the condensed phase relations in the systems Fo–Wo–FeO–Sil and Fo–Wo–Fe$_3$O$_4$–Sil under different P_{O_2} conditions. The systems are petrologically relevant in assessing the role of P_{O_2} on the crystallization trends of basalts.

The system Fo–Wo–FeO–Sil simulates low P_{O_2} conditions, and its phase relations are shown in Fig. 19a. There are prominent phase volumes of, particularly, olivines and pyroxenes with small volumes of silica polymorphs and wollastonite. The presence of wustite indicates low P_{O_2}. The important quaternary invariant points in the system are P, R (and S), representing continuous iron enrichment.

At low P_{O_2} the fractional crystallization of haplobasaltic liquid represented by point X (Fig. 19a) has the ability to produce high iron residues approaching the invariant point R, or in extreme cases even S, in composition. Thus the crystallization trends are clearly toward iron enrichment.

Figure 19b shows the relations in the system Fo–Wo–Mt–Sil, representing high P_{O_2} conditions shown by the presence of magnetite. It is apparent that magnetite is an important phase volume in addition to those of olivine, pyroxenes, wollastonite, and silica polymorphs. Magnetite volume cuts off and reduces the volumes of olivine and pyroxenes thus preventing the liquids from moving into a high-iron field. The important invariant points are the peritectic R and eutectics D and E. The univariant lines near E and those between R and E are not of much petrological interest.

Selected Mafic Silicate Systems with Iron Oxides

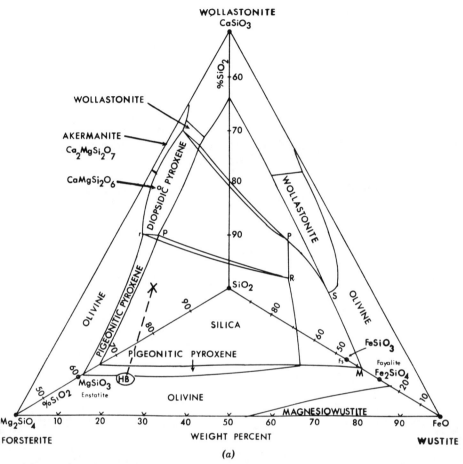

Figure 19 (a) Condensed equilibrium phase relations in the system forsterite–wollastonite–FeO–silica under low P_{O_2} conditions. (After Osborn, 1962. With permission of the *American Mineralogist*.) Solid lines $rR, pR, \ldots$ are univariant lines. The quaternary-invariant points P, R, and S reflect continuous iron enrichment.

At high P_{O_2} haplobasaltic liquid composition represented by point X (Fig. 19b) will tend to approach point D upon cooling and fractional crystallization. At D temperature and composition remain constant, and the liquid crystallizes to a final mixture of Pig + Di + Mt + Qtz solid phases defining a trend of silica enrichment. Therefore at high P_{O_2} the trend is toward forming silica-rich residues.

It is deduced from the above discussion that in silicate systems containing iron and approximating basaltic compositions, the liquid trend, upon fractional

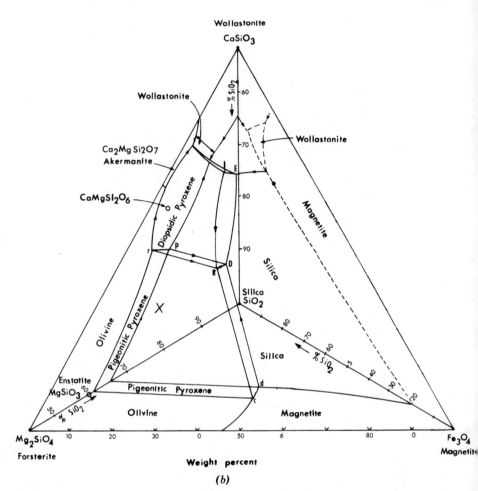

Figure 19 *(b)* Condensed phase relations in the system forsterite–wollastonite–magnetite–silica under P_{O_2} of 0.21 atm conditions. (After Osborn, 1962. With permission of the *American Mineralogist*.) Solid lines trending toward quaternary invariant points R and D are univariant lines which have three solid phases in equilibrium with the liquid.

Quaternary Point	Univariant Lines	Phases
R	rR	Ol + Pig + Di + Liq
(Ol + Pig + Di + Mt + Liq)	CR	Ol + Pig + Mt + Liq
	IR	Di + Ol + Mt + Liq
	RD	Pig + Di + Mt + Liq
D	dD	Pig + Mt + Qtz + Liq
(Pig + Di + Mt + Qtz + Liq)	pD	Di + Pig + Qtz + Liq
	ED	Di + Mt + Qtz + Liq
	RD	Pig + Di + Mt + Liq

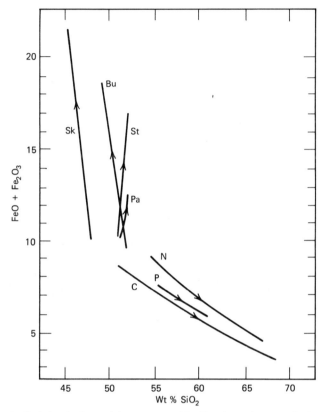

Figure 19 (c) Variation in the total iron content with silica for various petrological complexes. The divergence in the trend for the layered intrusions and volcanic complexes reflects the role of P_{O_2} in their crystallization. (After Osborn, 1962. With permission of the *American Mineralogist*.) Sk, Skaergaard, Greenland; Bu, Bushveld, South Africa; Pa, Palisade diabase sill, New Jersey; St, Stillwater, Montana; C, Cascade Series, U.S.A.; P, Paricutin Islands; N, Nockold's averages.

crystallization, is dependent on the prevalent P_{O_2} in the system. Hamilton and Anderson (1967) have reviewed the effect of P_{O_2} on the crystallization of basalts.

It is possible that iron enrichment observed in layered gabbroic (Fig. 19c) intrusions is a reflection of low P_{O_2} in the magma, whereas the orogenic andesite–dacite–rhyolite associations may suggest crystallization of basaltic magma under high P_{O_2} conditions (Osborn, 1963).

11.2 The System MgO–FeO–Fe$_2$O$_3$–SiO$_2$

The liquidus phase relations in this system are shown in Fig. 20a. The phase volumes of various silicate and oxide phases are depicted in the diagram. The

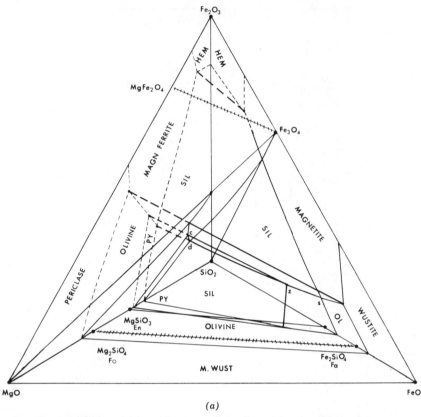

(a)

Figure 20 (a) Condensed phase equilibrium relations in the system MgO–FeO–Fe$_2$O$_3$–SiO$_2$ at 1 atm. (From Roeder and Osborn, 1966. With permission of the *American Journal of Science*.) The heavy solid lines are the univariant lines where three solid phases + Liq are in equilibrium. The dashed lines indicate estimated boundaries. Cross-hatches connect solid solution series. M. Wust, magnesio wustite; MAGN. FERRITE, magnesio ferrite (for explanation of the other abbreviations, see List of Abbreviations).

(b) Phase equilibrium relations in the plane MgO–Fe$_3$O$_4$–SiO$_2$ with fractional crystallization paths for three different P_{O_2}. All iron oxide is represented as Fe$_3$O$_4$. The light dashed lines represent relations at constant CO$_2$: CO = 32. Heavy solid lines are for P_{O_2} = 10$^{-0.7}$ atm and light solid lines for oxygen-buffered conditions of P_{O_2} > 10$^{-0.7}$ atm. Fractionation path for composition m is shown by the line m–n–o and that for m' by the curve m'–n'. (After Osborn, 1976. With permission of the International Congress on Thermal Waters.)

(c) Relationship between silica and iron-enrichment index as a function of varying P_{O_2} along the crystallization paths in the plane MgO–Fe$_3$O$_4$–SiO$_2$. (After Osborn, 1976. With permission of the International Congress on Thermal Waters.)

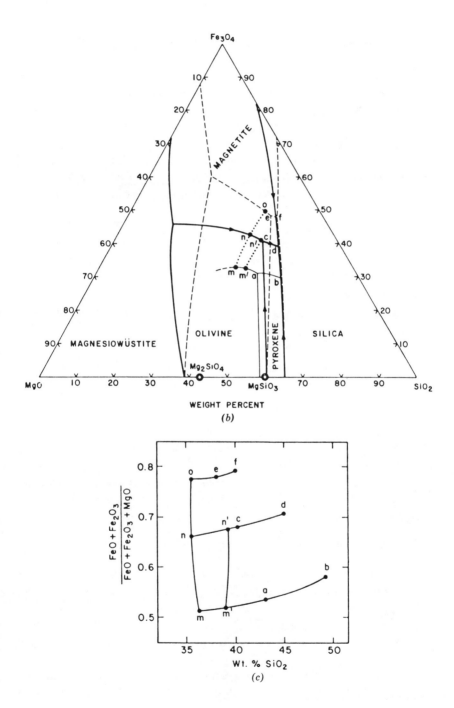

heavy solid lines are the univariant lines originating in the limiting ternary systems.

Petrologically, the plane $MgO-Fe_3O_4-SiO_2$ (Fig. 20b) is an important plane in the system, as it accommodates mineral phases common to basalts.

Therefore the system is pertinent to assess the role of P_{O_2} in the crystallization of basaltic and andesitic magmas. Osborn and Watson (1977) illustrated the applicability of the phase relations in the plane $MgO-Fe_3O_4-SiO_2$ to possible fractionation trends of a basaltic composition "m" (Fig. 20b).

The fractionation paths for a liquid of composition m, under different P_{O_2} conditions, can be demonstrated as follows with the help of Fig. 20b.

1. **Lowest P_{O_2} conditions.** Liquid m under these conditions will probably follow a course such as m–n–o–e–f. Along m–n–o, olivine separates and the iron-enrichment index (Fe/Fe + Mg) of the liquid increases. Along o–e, Ol + Mt separate, and the liquid shows silica enrichment along o–e–f. Thus along e–f, Px + Mt separate.

2. **Intermediate P_{O_2} conditions.** The fractionation path under these conditions is given by the curve m–n–n'–c–d, which characterizes a tendency toward silica enrichment relative to lowest P_{O_2} conditions considered above.

3. **High P_{O_2} conditions.** The fractionation path under high P_{O_2} is shown by the line m–m'–a–b. It is clearly an example of a trend toward the highest silica enrichment.

Figure 20c shows the silica-enrichment values for different fractionation trends described above, and it is apparent that increasing P_{O_2} favors silica enrichment similar to the proposal by Bowen (1928). Thus an andesite magma may be fractionally derived from a basaltic magma through simultaneous separation of magnetite and silicate phases. The presence of such phases in the groundmass of basalts gives credence to this inference (Osborn and Watson, 1977).

11.3 The System Forsterite–Anorthite–Magnetite–Silica

The above iron-bearing systems have a limitation to their broader applications since they lack feldspar as a component. So the system Fo–An–Mt–Sil is more appropriate to basaltic-andesitic compositions. The phase relations in this system (after Roeder, 1960) are shown in Fig. 21a and are relevant to describe the possible crystallization trends in basaltic rocks under conditions of varying P_{O_2}.

In this system at high P_{O_2} the magnetite volume cuts off the pyroxene and olivine volumes, similar to the cases discussed above. P–E–p–e is a prominent divariant surface of Px–An at about 50% An. The point P is a quaternary reaction point. The system has a eutectic, E, toward which liquids trend upon fractional crystallization. At E, Pl + Px + Mt + Qtz are stable. This is similar to the assemblage at the eutectic D in the system Fo–Wo–Fe_3O_4–Sil (Fig. 20b) characterizing a silica-enrichment trend and possible derivation of andesite from a basaltic parent as a function of P_{O_2} conditions.

Osborn and Arculus (1975) extended the studies to 10 kb in the temperature range of 1320–1420°C. Figure 21b shows the present status of phase relations at 10 kb. The P_{O_2} was maintained with magnetite–hematite (Mt–Hem) buffer. The changes in phase relations from 1 atm to 10 kb can be summarized as follows:

1 With increase in pressure, there is a shift in the location of Px–An surface toward the An apex. The Px–An surface at 10 kb is situated at 60% An rather than at 50% An as at 1 atm. Therefore the An volume contracts, while Px and Fe–Sp volumes expand. The implication is that fractional crystallization of Px–Sp in basalts at high pressure could push the residual liquid into the Px–An surface at low pressure—a trend toward Al_2O_3 enrichment. Thus anorthite or plagioclase could become a liquidus phase in andesites or calc-alkaline rocks. The melting studies on andesites (see Brown and Schairer, 1967) attest to this. Therefore a mechanism to form high Al_2O_3 derivatives from pressure-dependent fractional crystallization of basalts is demonstrable here.

2 The eutectic D at 10 kb replaces the reaction point P and eutectic E at 1 atm. At D, high Al–Opx + An + Mt + Trid are in equilibrium with liquid and gas.

3 Owing to lack of coexistence of An + Ol in this system at 10 kb, the univariant line Mt + Ol + Px, originating at *"a,"* is more stable than the two univariant lines kP and Pp' (Fig. 21b).

4 There is a temperature increase of 100°C for the different melting reactions.

5 Upon fractional crystallization at high pressures, basaltic liquids have the tendency to form higher Al_2O_3 derivatives than at 1 atm.

6 The phase boundary bD (Fig. 21b) is a magnetite (Fe–Sp) reaction curve. Along bD, as Px + An crystallize, magnetite is dissolved. Liquids that initially separate Fe–Sp + Px will stop separating magnetite as the liquids, through fractional crystallization, reach the curve bD. Magnetite will separate again only when the quaternary eutectic point D is reached. Therefore, depth at which fractional crystallization takes place is an important determinant of liquid trends.

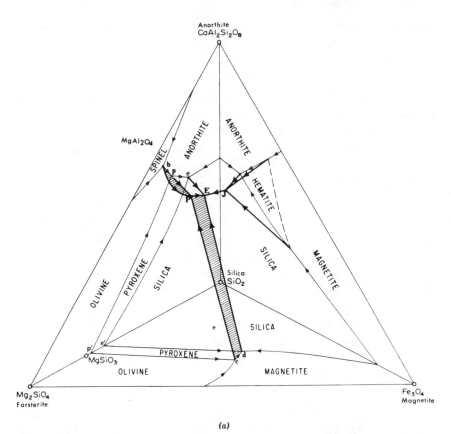

(a)

Figure 21 (a) Phase equilibrium relations in the system forsterite–anorthite–magnetite–silica at a P_{O_2} of 0.21 atm. Point E is a quaternary eutectic where An + Px + Mt + Qtz + Liq are in equilibrium. Arrows indicate direction of falling temperature. (After Roeder and Osborn, 1966. With permission of the *American Journal of Science*.)

Quaternary Point	Univariant Lines	Phases
P	cP	Ol + Mt + Px + Liq
(Ol + Px + An + Mt + Liq)	pP	Ol + Px + An + Liq
	bP	Ol + Mt + An + Liq
E	PE	Px + An + Mt + Liq
(Px + An + Mt + Qtz + Liq)	dE	Px + Mt + Qtz + Liq
	eE	Px + An + Qtz + Liq
	JE	An + Mt + Qtz + Liq

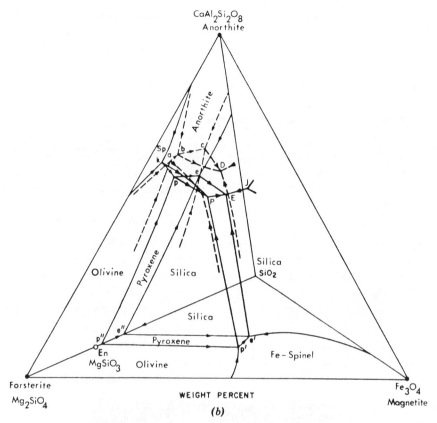

Figure 21 *(b)* Condensed phase equilibrum relations in the system forsterite–anorthite–magnetite–silica at 10 kb. The relations at 1 atm are shown in solid lines. Dashed lines are estimated curves at 10 kb under magnetite–hematite–buffered conditions. The quaternary eutectic D shifts toward the anorthite apex at higher pressure (cf. point E in Fig. 21a) and has Px + An + Mt + Qtz + Liq in equilibrium. (After Osborn and Arculus, 1975. With permission of the Carnegie Institution of Washington.)

Osborn and Arculus (1975, p. 507) suggest that "olivine basalt fractionally crystallizing at shallow depths will develop the calc-alkaline trend when magnetite begins to crystallize and magnetite will then continue as a liquidus phase."

However, fractional crystallization of basalts at deeper crustal levels may produce an andesite derivative liquid to reach the surface. In such cases plagioclase will be the primary phase (phenocryst) followed by some pyroxene and minor magnetite. Under near-surface fractional crystallization, there is some concomitant enrichment in iron in the late silica-rich derivatives until magnetite is crystallizing again. Thus magnetite, as phenocrysts, may be absent (cf. Cascade series; Carmichael and Nicholls, 1967) if initial fractionation occurred at high pressures.

The phase relations in the system Fo–An–Fe$_3$O$_4$–Sil are useful in explaining the role of P_{O_2} and crystal-liquid processes in the formation of basalt–andesite–dacite–rhyolite series of many orogenic areas.

11.4 Silicate Systems with Chromium Oxide

Chromium is an important constituent of spinels found in basic and ultrabasic igneous rocks. Chrome spinels are stable with olivine and pyroxenes and separate rather early in the crystallization history of a basaltic magma containing as low as 0.1% Cr$_2$O$_3$.

Chromites (chrome spinels) form almost monomineralic layers (chromitites) in many layered intrusions, for example, Muskox, Canada; Bushveld, South Africa; Karnataka, India; and Alpine peridotite and ophiolite complexes.

In order to explain the origin of chrome spinels, their associations with olivine–pyroxene assemblages, their compositional variability and their pressure–temperature-dependent crystallization, many studies have recently begun on silicate and Fe–Cr–oxide systems. For a brief discussion of the system Di–Fo–An–Sil–MgCr$_2$O$_4$ see Section 6.1. The reader is referred to the investigations by Arculus et al. (1974), Arculus (1974), Arculus and Osborn (1975), Dickey et al. (1971), Ulmer (1969), Keith (1954) and Irvine (1977).

A specimen of a layered intrusion with chromitite (Cr) and dunite-harzburgite (D-Hz) layers from Archean ultramafic complex near Dodkanya, Karnataka, India. (Sample courtesy of Dr. Sri Kantappa, University of Mysore, India.)

CHAPTER 3
Anhydrous Silicate Systems Related to the Crystallization of Residual Alkali Magmas

1 NATURE OF THE RESIDUAL LIQUIDS

It is clearly seen from the phase relations in the systems (see Figs. 1–8, 14, and 15–21) that the low temperature liquids are those rich in alkalis, alumina, and silica. They are poor in lime, magnesia, and ferric iron, if iron is part of the components of the systems. Such low temperature liquids may be considered as residual liquids obtained through fractional crystallization of mafic melts. With crystallization, this is an important aspect of the trend in liquid composition. Residual liquids therefore must correspond to peritectics or eutectics as those are the low melting compositions in the multicomponent systems. It is appropriate now to consider the phase relations in systems containing alkalis, alumina, and silica to investigate the crystallization behavior of the residual liquids and their interrelations with mafic counterparts.

2 THE SYSTEM NEPHELINE–KALSILITE–SILICA: PETROGENY'S RESIDUA SYSTEM

Bowen (1935, 1937), on the basis of experimental investigations of various silicate systems, advocated that residual liquids derived from fractional crystallization of natural mafic melts must tend to approach, in composition, the system Ne–Ks–Sil. Therefore he proposed that the system Ne–Ks–Sil be termed the *Petrogeny's Residua System*.

The equilibrium relations for the system Ne–Ks–Sil are shown in Fig. 22a. It has prominent phase areas of alkali feldspars, Na–K nephelines, kalsilite, carnegieite, leucite, and tridymite-cristobalite. The arrangement of isothermal

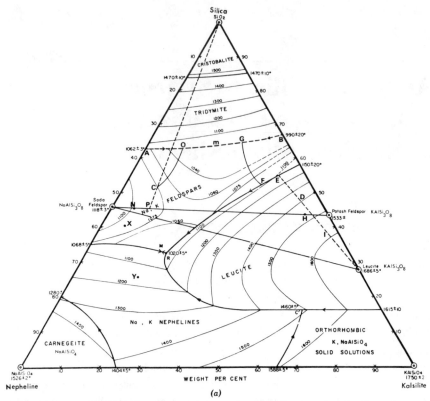

Figure 22 *(a)* Equilibrium phase relations in the Petrogeny's Residua System nepheline–kalsilite–silica with crystallization paths for selected compositions. (Modified from Bowen, 1937, by Schairer, 1950. With permission of the *Journal of Geology* and McGraw-Hill, Inc., New York.)

lines shows an interesting thermal surface for the system. A few important features of the system are noted below.

1. Albite and orthoclase are two important intermediate compounds on the Ne–Sil and Ks–Sil compositional lines. In the middle of the diagram (Fig.22a) they form the Ab–Or or the alkali feldspar join which divides the system into two parts. The portion Ab–Or–Sil pertains to the silica-oversaturated compositions such as granites (rhyolites) and is referred to as the "granite system."

 The portion Ab–Ne–Ks–Or is pertinent to the silica-undersaturated compositions, for example, nepheline syenites (phonolites) and is termed the "nepheline syenite" or "phonolite system."

The silica-saturated compositions such as syenite (trachytes) are represented by the Ab–Or join itself.

2   In terms of thermal relations, the system has two temperature minima. One at ~ 950°C on the quartz–alkali feldspar phase boundary is called the "granite minimum" *(m)*. The other (at <1020°C) on the nepheline–alkali feldspar phase boundary is termed the "phonolite minimum" *(M)*.

3   The two minima are separated by a temperature crest at 1063°C on the alkali feldspar join. Therefore the join is a "thermal divide." However, incongruent

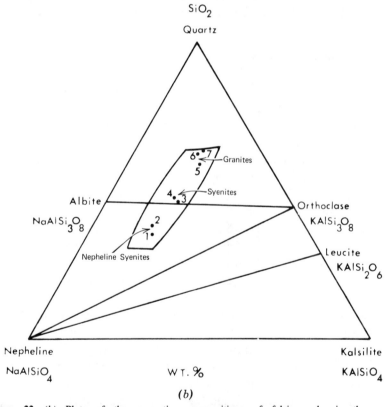

Figure 22 *(b)* Plots of the normative compositions of felsic rocks in the system nepheline–kalsilite–silica. Note the rocks plot in the thermal valley of the diagram separating the fields of granites (rhyolites), syenites (trachytes), and nepheline syenites (phonolites). (After Bowen, 1937. With permission of the *American Journal of Science*.) 1. tinguaites   2. phonolites   3. alkali syenites   4. alkali trachytes   5. pantellerites   6. granites   7. rhyolites

melting of orthoclase decreases the effectiveness of the "thermal divide" in the leucite field.
4 There is no eutectic in the system but a peritectic R (Fig. 22a) involving leucite reaction. The phase boundary SR is Lc + liq → Alk-Fels reaction curve over most of its course.
5 Upon crystallization, the two minima are the goals for appropriate liquids to reach.
6 The extensive solid solution in alkali feldspar and nepheline permits a great flexibility in the fractionation process.

Let us now consider equilibrium crystallization of compositions in the system with possible effects of fractionation and the nature of the thermal divide.

2.1 Crystallization Trends

The incongruent melting of orthoclase and the solid solution nature of the phases permit discussion of only general trends of crystallization.

A melt of composition C (Fig. 22a) lying in the Ab–Or–Sil portion, at appropriate temperature, crystallizes an alkali feldspar of composition N. With further cooling, the melt takes a unique curved course, such as CO, to the feldspar–silica boundary AB. Along this boundary alkali feldspar and tridymite crystallize together, and the liquid phase is eliminated. The final crystalline product is Alk-Fels + Trid. The composition of the final alkali feldspar is obtained by drawing a tie line (Sil–C–P; Fig. 22a) from the silica apex through the original melt C to intersect the alkali feldspar join at P. Therefore the point P gives the composition of the alkali feldspar.

The composition D in the leucite field (Fig. 22a) but above the alkali feldspar join will, upon crystallization of leucite, move directly away from the leucite point along the line DE to the Lc–Alk–Fels phase boundary. Along EF leucite dissolves to form alkali feldspar. As all leucite is changed over to alkali feldspar the liquid takes a curved course FG to the Alk-Fels–Sil phase boundary. The final crystalline mixture is Alk-Fels + Trid as the last of the liquid is used up. The composition of the final alkali feldspar is given by point H.

It is important to note that, although leucite is the first phase to form, the silica content of the melt is greater than the stoichiometeric value for the stability of leucite (cf. systems Lc–Sil, Schairer and Bowen, 1935; Fo–Sil, Fig. 5). Moreover, D lies in the Ab-Or-Sil compositional triangle.

For all compositions that have albite as the liquidus phase the appearance of leucite will be inhibited.

For compositions such as X or Y lying in the feldspar and nepheline fields, respectively (Fig. 22a), crystallization will terminate at the phonolite minimum M to a final solid mixture of Ne + Alk-Fels. Nepheline and alkali feldspars will, of course, continuously change composition until crystallization ceases.

A melt of composition I (Fig. 22a) in the leucite field, but in the silica-undersaturated portion of the system, crystallizes leucite as the first phase. With cooling, the melt moves along the line IDE as leucite separates. At E, K-feldspar crystallizes. It also forms by leucite-liquid reaction as the liquid moves down the line ER. At the invariant point R, temperature and composition remain constant until all the liquid is used up. The final crystalline product is a mixture of Ne + Alk-Fels + Lc.

However, for compositions in the leucite field the final assemblage may be quite varied, compared to the simple case just discussed, as leucite is a solid solution component.

The above discussion explains some important features of alkaline rocks.

1. The occurrence of leucite in rapidly cooled silica-rich groundmass due to limited stability of leucite with silica-rich liquids, for example, compositions in the leucite field above the alkali feldspar join.
2. The rarity or common absence of leucite in plutonic rocks (for further discussion see Chapter 4, section 8).
3. Pseudoleucites (Ne + Alk-Fels mixtures) form from leucite reaction at the point R.
4. The complex solid solution nature of nephelines and the dominance of sodic nephelines in phonolites/nepheline syenites.
5. Under equilibrium crystallization, liquids below the alkali feldspar join cannot produce quartz-bearing assemblages. Similarly those above the join cannot give nepheline-bearing assemblages.

The general separation in space and time of nepheline- and quartz-bearing alkaline rocks attests to the possibility that a "thermal divide" similar to alkali feldspar join may also persist in nature. This suggests that a fundamental process must operate during the very early stages of crystallization of the parental magmas to dictate the possible chemical lineage or liquid line of descent controlled by fractionation.

However, the alkali feldspar "thermal divide" may be penetrated by Fe^{3+}–Al^{3+} substitution in Na-feldspars (see Bailey and Schairer, 1966), volatile transfer of alkalis and silica (Smyth 1927), and extreme enrichment or fractionation of leucite or addition of lime. Alkaline rocks of the Stettin and Wausau area, Wisconsin (Sood, et al., 1980) and Kangerdlugssuaq intrusion, East Greenland (Wager, 1965) may be examples of such a penetration.

Rocks formed from the crystallization of residual magmas should plot reasonably in or close to the low temperature belt in the system, if they are true derivatives. Figure 22b shows the plots of phonolites, trachytes, rhyolites, syenites, granites, and so on, which lie in the thermal valley. Some offsetting is caused due to the presence of plagioclase feldspar, but the inference is clear that crystal–liquid process exerts a dominant role in their formation

(Certain aspects on the effect of water vapor pressure on this system are described in Chapter 4 section 8.)

3 THE SYSTEM DIOPSIDE–NEPHELINE–KALSILITE–SILICA: THE ALKALI ROCK TETRAHEDRON

The system Ne–Ks–Sil (Fig. 22a) is the plane of residual alkaline liquids formed through fractionation of mafic magmas (Bowen, 1937). However, the presence of appreciable amounts of pyroxenes in alkaline rocks necessitates its expansion. The system Di–Ne–Ks–Sil, containing an iron-free representative of pyroxenes, is a step toward providing a better physicochemical framework in understanding the crystallization behavior and interrelationships of both sodic and potassic alkaline melts.

Sood et al. (1970) have studied phase equilibrium relations in portions of the system Di–Ne–Ks–Sil. The system is herein referred to as the Alkali Rock Tetrahedron (Fig. 23) and may be considered a potash analog of the simple Basalt Tetrahedron of Yoder and Tilley (1962) discussed previously. There are many parallels with the simple Basalt Tetrahedron thus permitting assessment of interrelations of basic and felsic undersaturated alkaline rocks with both sodic and potassic affinities. However, due to the complex thermal relations and extensive solid solution among the phases, the planes Di–Ab–Lc and Di–Ab–Or (Fig. 23) can, only in a restricted sense, be designated as a "plane of critical silica undersaturation" and a "plane of silica saturation," respectively.

The Alkali Rock Tetrahedron includes the major iron-free normative and modal mineral phases of most alkaline rocks. Iron, although an important con-

The System Diopside-Nepheline-Kalsilite-Silica

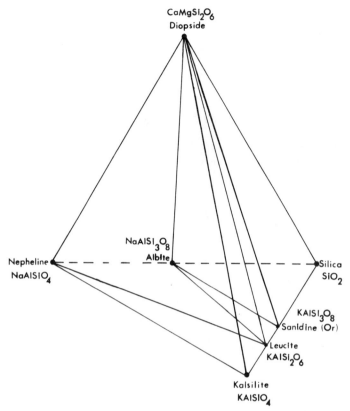

Figure 23 Schematic representation of the Alkali Rock Tetrahedron diopside–nepheline–kalsilite–silica showing planes of silica saturation (Di–Ab–Or) and undersaturation (Di–Ab–Lc). (Modified from Sood et al., 1970. With permission of the *Canadian Mineralogist*.)

stituent mainly in pyroxenes (see Tyler and King, 1967), is excluded to reduce the complexity of the system. However, Tilley (1957), Yagi (1962), Bailey and Schairer (1966), and Nolan (1966) have pointed out the modifications that an iron-bearing pyroxene (principally acmite) introduces in the crystallization trends of synthetic and natural alkaline liquids. A part of the iron effect is discussed in section 6 of this chapter.

3.1 Characteristics of the Alkali Rock Tetrahedron

Figure 24a,b represent a generalized and enlarged view of the tetrahedron Di–Ne–Ks–Sil, where diopside has been replaced by a clinopyroxene for broader

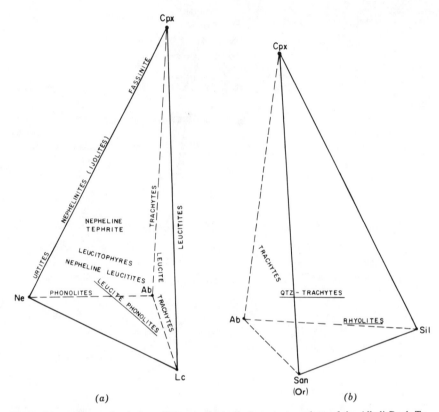

Figure 24 (a) Generalized view of the petrologically important portions of the Alkali Rock Tetrahedron with plots of sodic and potassic feldspathoidal rocks in the silica undersaturated portion Cpx–Ne–Lc–Ab. (b) Silica saturated and oversaturated alkaline rocks in the portion Cpx–Ab–San–Qtz.

representation of alkaline rocks. Included in Fig. 24 are the names of rocks whose modal (and normative) compositions lie in the tetrahedron. The modal information for rock names has been taken from Johannsen (1939). Despite its limitations with respect to lack of iron, the tetrahedron accounts for a large variety of alkaline rocks (see Table VIII), both sodic and potassic.

A few important points related to the tetrahedron are summarized below.

1 The portion Cpx–Ab–Or–Sil (Fig. 24b) encloses the silica-saturated and -oversaturated compositions representing simplified trachytes and rhyolites, respectively (or their plutonic equivalents). However, some leucite may ap-

Table VIII Normative Classification of Alkaline Rocks

1. Rhyolites	Silica-oversaturated	Quartz-normative
2. Trachytes	Silica-saturated	Feldspar- and/or quartz-normative
3. Leucite trachytes Leucite phonolites Phonolites	Silica-undersaturated	Nepheline-normative
4. Leucitophyres Leucitites Nepheline leucitites	Silica-undersaturated	Nepheline- and possibly leucite-normative
5. Leucite nephelinites Olivine nephelinites Nephelinites Kalsilite nephelinites	Highly silica-undersaturated	Nepheline-, leucite- and/or kalsilite- and olivine-normative

pear in the mode along with quartz in silica-saturated rocks as a result of the incongruent melting of orthoclase to leucite under volcanic conditions. Such examples have been described by Hussak (1900).

2 In the silica undersaturated part of the tetrahedron compositions near the join Di–Ab–Lc (Fig. 23) are characterized by silica deficiency as shown by nepheline* in their norms. Alkaline lavas, with compositions near this join, contain modal leucite rather than nepheline, although frequently pseudoleucite may be present as a result of leucite–liquid reaction. Examples of this type are described by Cooke and Moorhouse (1969).

3 In the more silica undersaturated portion of the tetrahedron (Fig. 24a) compositions containing both modal and normative nepheline, characterizing phonolitic type rocks, are encountered. Leucite, though present modally, is absent in the norms. It is only near the plane Di–Ne–Lc (Figs. 23 and 24a)

---

*In the CIPW scheme of norm calculation for alkaline (or other) compositions, the appearance of nepheline in the norm defines the silica-undersaturation index of the composition. In doing so albite is preferentially desilicated, instead of orthoclase, to satisfy the silica requirement, making albite and leucite incompatible in the norm. Therefore, due to this preferential desilication of albite rather than orthoclase, leucite may not appear at all in the norms of many highly silica-undersaturated potassic compositions (e.g., leucitites), although it may be abundant in the mode. Therefore the norm calculation mitigates against leucite formation and consequently does not depict clearly the leucitic nature of the rock (Holmes, 1930).

that leucite and nepheline may appear in the norms as well as in the modes. Compositions near this plane may represent assemblages corresponding to simplified nephelinites, nepheline leucitites, leucitophyres, and so on.

4 Highly undersaturated compositions appear near the Di–Ne–Ks face. Such compositions carry nepheline, leucite, and/or kalsilite in the norms and may yield essentially feldspar-free assemblages corresponding to simple olivine nephelinites and certain kalsilite-bearing lavas (kalsilite nephelinites).

3.2 Phase Relations in the System Di–Ne–Ks–Sil

Consideration of the phase relations in the tetrahedron Di–Ne–Ks–Sil requires a brief description of its limiting ternary systems, Di–Ne–Ks, Di–Ne–Sil, Ne–Ks–Sil, and Di–Ks–Sil.

The relations in the system Ne–Ks–Sil have just been presented in the previous section and that for Di–Ne–Sil in Chapter 2, Section 7. No data are available for the system Di–Ne–Ks, but it may be important to fully define the relations in the tetrahedron.

The system Di–Lc–Sil representing a portion of the system Di–Ks–Sil was studied by Schairer and Bowen (1938), and the phase relations are shown in Fig. 8b. In this system the join Di–Or does not form an equilibrium thermal

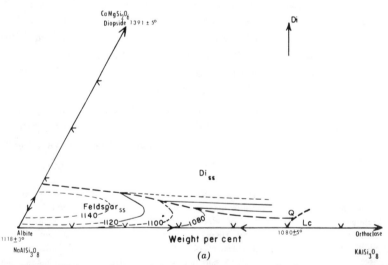

Figure 25 Equilibrium phase relations in the joins of the system diopside–nepheline–kalsilite–silica. (a) The join diopside–albite–orthoclase. (After Morse, 1968a. With permission of the Carnegie Institution of Washington.)

divide due to the incongruent melting of orthoclase. It has two invariant points, the reaction point E at about 1100°C and the eutectic F at about 990°C (Fig. 8b). Diopside has a large phase area and is the first phase to appear even for compositions containing only about 2.0 wt.% of the mineral.

In addition to data on the limiting faces of the tetrahedron, phase relations are also known for joins Di–Ab–Or (Morse, 1968), Di–Ab–Lc, and Di–Ne–San (Sood et al., 1970).

The join Di–Ab–Or (Fig. 25a) is pseudoternary due to the incongruent melting of orthoclase and ternary feldspar solid solution. It has one piercing point Q at ~1060°C.

The join Di–Ab–Lc (Fig. 25b) is also pseudoternary due to solid solution among the phases and has a piercing point P at 1040°C. Thus liquids migrate out of the join toward a quaternary minimum.

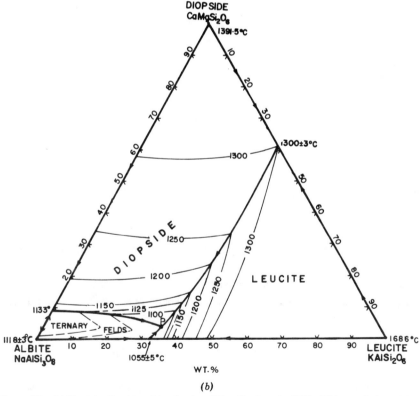

Figure 25 (b) The join diopside–albite–leucite. (After Sood et al., 1970. With permission of the *Canadian Mineralogist*.)

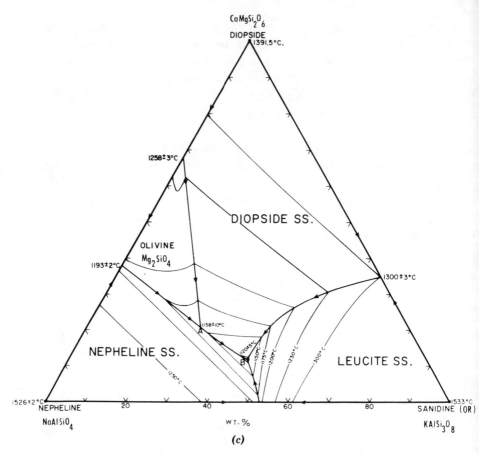

Figure 25 (c) The join diopside–nepheline–sanidine (After Platt and Edgar, 1972. With permission of the *Journal of Geology*.)

The join Di–Ne–San (Fig. 25c) is also pseudoternary due to the incongruent melting of sanidine and the appearance of forsterite field on the liquidus surface. It has two piercing points A and B with temperatures of 1158°C and 1120°C, respectively, the former being a reaction point where forsterite reacts with liquid-producing pyroxene (see Platt and Edgar, 1972).

3.3 Flow Sheet with Univariant and Invariant Relations

All the information on the limiting ternary systems presented above is combined to depict the relations in the tetrahedron (Fig. 26) where Ne–Ks–Sil forms the

The System Diopside-Nepheline-Kalsilite-Silica

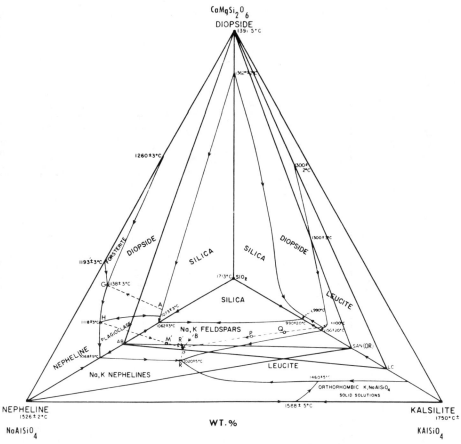

Figure 26 Condensed phase relations in the silica undersaturated part of the system diopside–nepheline–kalsilite–silica. (After Sood et al., 1970.) The dashed lines are the univariant lines. Letters show the location of the piercing points. The arrows indicate direction of falling temperatures. Point R' is a quaternary peritectic at ~5.0 wt. % diopside. (With permission of the *Canadian Mineralogist*.)

base, Di–Ne–Ks the front face and Di–Ne–Sil and Di–Ks–Sil the other two faces. The positions of the three joins Di–Ab–Lc, Di–Ab–Or, and Di–Ne–San (Or) are also shown. Figure 26 shows the phase volumes of silica polymorphs, forsterite, feldspars,* nephelines, leucite, and diopside.

The data on the piercing and the invariant points in the joins and limiting systems, respectively, are listed in Table IX. In discussing the univariant and

---
*The term feldspar (Fels) in this section refers to a ternary feldspar of undetermined composition.

Table IX Piercing Points of Univariant Lines in the Three Joins and the Limiting Ternary Systems of the Tetrahedron Di–Ne–Ks–Sil

Number	Join or System	Piercing Point	Three Solid Phases	Liquid Composition	Temp. (°C)	Reference
1	Di–Ab–Lc (Fig. 25b)	P	Di + Lc + Fels	$Di_4Ab_{62}Lc_{34}$	1040 ± 10	Sood et al., 1970
2	Di–Ab–Or (Fig. 25a)	Q	Di + Lc + Fels	$Di_3Ab_{48}Or_{49}$	~1060	Morse, 1968
3	Di–Ne–San (Fig. 25c)	B	Ne + Lc + Di	$Ne_{44.5}San_{43.5}Di_{12}$	1120 ± 5	Sood et al., 1970
4	Di–Ne–San (Fig. 25c)	A	Di + Ne + Fo	$Ne_{51}Di_{21}San_{28}$	1158 ± 10	Sood et al., 1970
5	Di–Lc–Sil (Fig. 8b)	E	Di + Lc + K–Fels	$Di_2Lc_{56}Sil_{42}$	1100 ± 10	Schairer and Bowen, 1938
6	Di–Lc–Sil (Fig. 8b)	F	Di + K–Fels + Trid	$Di_2Lc_{44}Sil_{54}$	990 ± 10	Schairer and Bowen, 1938
7	Di–Ne–Sil (Fig. 14b)	H	Di + Ne + Pl	$Di_{15}Ne_{65}Sil_{20}$	1118 ± 3	Schairer and Yoder, 1960a
8	Di–Ne–Sil (Fig. 14b)	G	Di + Fo + Ne	$Di_{28}Ne_{59}Sil_{13}$	1138 ± 3	Schairer and Yoder, 1960a
9	Di–Ne–Sil (Fig. 14b)	I	Di + Pl + Trid	$Di_3Ne_{36}Sil_{61}$	1073 ± 3	Schairer and Yoder, 1960a
10	Ne–Ks–Sil (Fig. 22a)	R	Ne + Lc + Fels	$Ne_{47}Ks_{21}Sil_{36}$	1020 ± 5	Schairer, 1957
11	Ne–Ks–Sil (Fig. 22a)	C'	Ne–Lc–Ks	$Ne_{20.5}Ks_{63.5}Sil_{16}$	1460 ± 5	Schairer, 1957

invariant characteristics of the system, attention is focused on the silica-undersaturated portion Di–Ab–Ne–Ks–Or (Fig. 26). The portion around the invariant points is enlarged in (Fig. 27) to show the details. The various univariant lines, the invariant points, and the volumes in which they are located are given in Table X and used in the construction of the flow sheet shown in Fig. 28.

Figure 28 is totally schematic. However, the positions of the joins and systems used in the construction of the flow sheet are shown. For any given liquid composition, such a diagram indicates the quaternary invariant point toward which the liquid proceeds during crystallization, and the quaternary minimum at which it will completely crystallize.

Along each of these univariant lines three solid phases are in equilibrium with the liquid. However, due to the crystallizing phases being solid solutions, the compositions, *sensu stricto,* do not lie on the univariant lines.* The letters and temperatures at the beginning of the lines refer to the ternary invariant (or piercing) points and their temperatures, respectively (as shown in Figs. 26 and 28 and given in Tables IX and X). The arrows on the lines indicate the direction of falling temperatures.

The following are some observations on the phase relations in the system Di–Ne–Ks–Sil.

1 The join Di–Ab–San (Fig. 23) subdivides the system Di–Ne–Ks–Sil into silica-undersaturated and oversaturated portions. It does not act as a true compositional plane due to the incongruent melting of sanidine (orthoclase) and extensive solid solution among the feldspars. However, the temperature of the beginning of melting observed by Morse (1968) for compositions in this join is 1020°C, whereas the temperature of the beginning of melting in the join Di–Ab–Lc is 970°C. Therefore, the join Di–Ab–San acts as an "effective barrier" under normal crystallization processes, for silica-undersaturated compositions. Sufficient concentrations of leucite for compositions in the leucite field may produce a liquid crystallizing Di + Fels + Qtz at some invariant point in the Di–Ab–Or–Sil volume (Fig. 26). Thus it is possible for certain undersaturated liquids to pass through this plane of silica saturation, by early fractionation of leucite, to yield oversaturated residual liquids.

2 There is no "true" maximum on the univariant lines in the undersaturated part of the tetrahedron Di–Ne–Ks–Sil; however, a minimum on the line HR' (Fig. 28) has been postulated.

*For convenience and clarity, the term *univariant line* will be used in the remainder of this discussion.

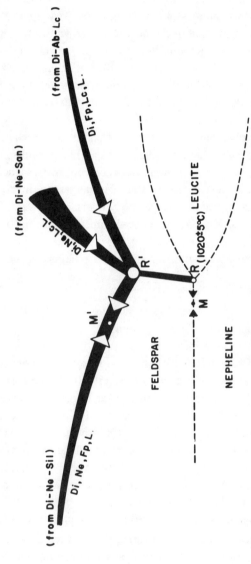

Figure 27 Exploded view around the quaternary peritectic R' of Fig. 26. The univariant lines are shown as solid lines. The systems and joins in which they originate are given in parentheses. The three solid phases in equilibrium with liquid are listed along the univariant lines. The temperature of R' is estimated at 1000 ± 20°C. M' is the minimum where the liquid crystallizes to only three solid phases due to solid solution. (After Sood et al., 1970. With permission of the *Canadian Mineralogist*.)

Table X Univariant Lines and Invariant Points Within the Various Volumes of the Tetrahedron Di–Ne–Ks–Sil[a]

The volume Di–Ab–Ne–Lc–San (Figs. 23 and 26)

Univariant Lines (Figs. 26 and 28)
 QPR'   Di + Lc + Fels + Liq
 HR'   Di + Ne + Fels + Liq
 BR'   Di + Ne + Lc + Liq
 RR'   Ne + Lc + Fels + Liq
Invariant Points (Quaternary)
 R'   Di + Ne + Lc + Fels + Liq
 M'[b] Di + Ne + Fels + Liq

The volume Di–Ne–Ks–Lc (Fig. 23)

Univariant Lines (Fig. 28)
 I'C   Ks + Ne + Lc + Liq
 I'J   Di + Ks + Lc + Liq
 I'K   Di + Ne + Ks + Liq
 I'B   Di + Ne + Lc + Liq
Invariant Point (Quaternary)
 I'   Di + Ne + Ks + Lc + Liq

[a] From Sood et al. (1970).
[b] M' is not a "true" invariant point but represents a postulated minimum in the system.

3 The forsterite volume near the Di–Ne join (Fig. 26) is an extension of the same from the Basalt Tetrahedron. From liquids in this volume melilite crystallizes as a subliquidus phase in compositions containing 10 wt. % K-feldspar but is absent in compositions with more than 10 wt. % of K-feldspar (also see Platt and Edgar, 1972).

However, at lower temperatures, melilite and forsterite disappear by reaction with the liquid. At low temperatures nepheline and pyroxene are stable assemblages, suggesting nephelinites to be low temperature equivalents of olivine-melilite rocks.

The presence of the forsterite volume and the independent nature of some of the oxides is indicative that the system is not truly quaternary. It could adequately be represented by $CaO-MgO-Na_2O-K_2O-Al_2O_3-SiO_2$. Due to the limited variation, it is considered a quaternary.

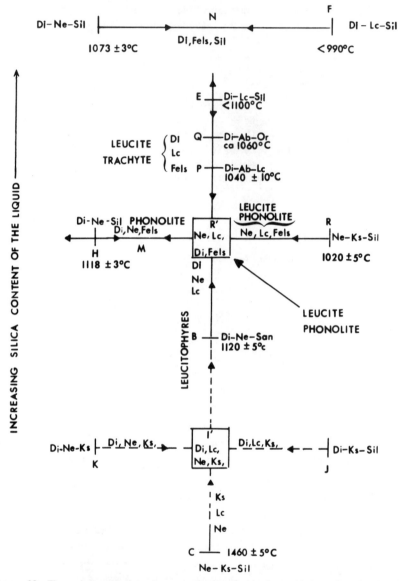

Figure 28 Flow sheet showing the univariant and invariant relations for the system diopside–nepheline–kalsilite–silica. Arrows indicate direction of falling temperatures. The three solid phases coexisting with liquid are shown along the univariant lines with possible rocks names corresponding to the assemblage. The dashed lines are not fully determined. (After Sood et al., 1970. With permission of the *Canadian Mineralogist*.)

4 Sood et al. (1970), from their studies, located one quaternary invariant point and postulated two more. Certain characteristics of the quaternary invariant points are described below.

a) The quaternary invariant point R' is located in the volume Di–Ab–Ne–Lc–San (Fig. 26), and the arrangement around this point is shown in an enlarged view in Fig. 27. The temperature of the point R' is probably less than 1040°C (point P, Fig. 28) and even less than 1020°C (point R, Fig. 22a), as the CaO content of the solid solution phases, nepheline and feldspar, is suspected to be small. Moreover, the points R, R', and M' are so close in composition (Fig. 26) that it is very difficult to delineate accurately the temperature differences between them. The viscous nature of liquids in this region, resulting in sluggish crystallization, is another added difficulty. Sood et al. (1970) estimate the composition of R' to be that of R (Fig. 20) with 5 wt. % diopside.

If the behavior of leucite in nature and in the limiting ternary systems is considered, and the fact that ER' (over most of its length) and RR' (Fig. 28) are reaction lines, then leucite is the phase that disappears by reaction at R' according to the equation

$$2KAlSi_2O_6 + NaAlSi_3O_8 \rightleftarrows 2KAlSi_3O_8 + NaAlSiO_4$$
$$\text{Lc} \qquad \text{Ab} \qquad \text{Or} \qquad \text{Ne}$$
(in melt)

Therefore, the quaternary invariant point R' is a reaction point. All compositions, within the volume Di–Ab–Ne–Lc–San (Fig. 26), crystallizing leucite must proceed to this point.

b) Point M' (Figs. 27 and 28) is the minimum on the liquidus in the volume Di–Ab–Ne–Lc–San (Fig. 26) postulated from the peritectic relations at the point R' and the appearance of Di + Ne + Fels as end products for certain compositions in the volume. The point M' is the ultimate goal for all liquids orginating in the silica-undersaturated portion of the system (Fig. 26), which will be facilitated by extreme fractionation. The temperature of this minimum is probably less than 1000°C as the temperature of beginning of melting for compositions in the join Di–Ab–Lc is 970°C. In composition it is close to the point R'. Due to the solid solutions and the solid–liquid reaction relation between the crystallizing phases, the last liquid crystallizes only three solid phases.

c) The point I' (Fig. 28) is the invariant point for highly silica-undersaturated K-rich liquids in the volume Di–Ne–Ks–Lc (Fig. 26). The temperature of this point must be less than 1460°C (point C, Fig. 28) and greater than 1120 ± 5°C (point B, Fig. 27). At I' the last of the kalsilite disappears, either by reaction or by inversion. The residual liquids then follow the line $I'BR'$ into the volume Di–Ab–Ne–Lc–San (Fig. 26) to point R'.

The flow sheet in Fig. 28 is useful in charting the course of liquids after the disappearance of olivine and melilite phases. Liquids may terminate crystallization either at the minimum M' after passing through R' or at I' (Fig. 28) for highly silica-undersaturated compositions. The paths will be dictated by the initial composition. Fractionation will increase the tendency to reach the invariant points, but many liquids may crystallize as they approach these invariant points.

3.4 Liquid Trends

Consideration of the phase relations in the system Di–Ne–Ks–Sil, suggests that the liquids in the early stages of crystallization may form two distinct liquid trends, one sodic and the other potassic. In the later stages the two trends converge to terminate final crystallization at the minimum M' (Fig. 26) to give Di + Ne + Fels corresponding to a simple phonolite.

Nepheline (Sodic) Trend

For compositions near the forsterite volume (Fig. 26) olivine and melilite are stable at high temperatures. At lower temperatures olivine and melilite change over to pyroxene and nepheline by reaction. This suggests the possibility that a nepheline-pyroxene rock may be a low temperature equivalent of the olivine-melilite rock.

Fractional crystallization of olivine and/or melilite from liquid compositions on the Di–Ne surface but close to the forsterite volume may, according to Platt and Edgar (1972), result in the following possible liquid trends.

1 Olivine nephelinite–nephelinite–phonolite
2 Olivine melilitite–melilite nephelinite–phonolite
3 Olivine melilite nephelinite–melilite nephelinite–nephelinite–phonolite

It is apparent that such liquid compositions will leave the tetrahedron and reenter at some point between H and R' (Figs. 26 and 28) after separation of

The System Diopside-Nepheline-Kalsilite-Silica

olivine and/or melilite. Therefore final crystallization will be on the line HR' (Fig. 28) to a simple phonolite.

Leucite (Potassic) Trend

The liquid compositions in the leucite volume enclosed by the planes Di–Ab–Lc and Di–Ne–Lc have, through fractionation, the potentials of producing varied assemblages defining, in a simplified manner, dominantly potassic trends. These assemblages are listed in Table XI and translated to corresponding rock names in Fig. 29. Some possible trends are:

1. Liquids on the Di–Ne surface, upon fractionation of olivine and/or melilite, reenter the tetrahedron at point B (Figs. 26 and 28). Such liquids, after their passage to the point R' (Fig. 28), may give rise to the trend olivine nephelinite–leucitophyre–leucite phonolite–phonolite (Fig. 29). This trend represents a merger of the sodic and potassic trends and is elaborated on later.

2. Liquids on the Di–Lc surface, through successive fractionation of diopside and leucite, reach the point R' and possibly finally the minimum M' on the line HR' (Fig. 28). Such a fractionation sequence has the possibility of giving the liquid trends shown in Fig. 29 and listed on page 98.

Table XI Various Possible Assemblages[a] from Liquid Compositions Near the Di–Ne and Di–Lc Surfaces

For compositions near the Di–Ne surface			
1	2		
Fo + Di + Ne	Fo + Di + Ne		
Di + Ne + Pl	Di + Ne + Lc		
Di + Ne + Fels	Di + Ne + Lc + Fels		
	Di + Ne + Fels		

For compositions near the Di–Lc surface			
1	2	3	4
Di + Lc	Di + Lc	Di + Lc	Di + Lc
Di + Lc + Fels	Di + Lc + Fels	Di + Lc + Ne	Di + Fels
Di + Lc + Fels + Ne	Di + Fels + Ne	Di + Ne + Lc + Fels	Di + Fels + Qtz
Di + Fels + Ne	—	Di + Fels + Ne	

[a] See Fig. 29 for rock associations.

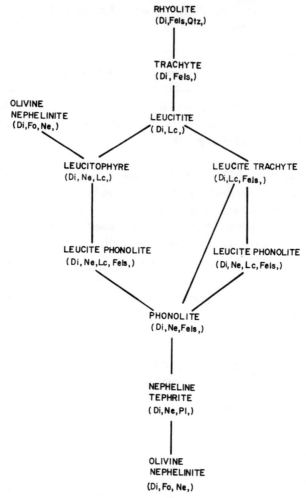

Figure 29 Flow sheet showing liquid trends, possible through fractionation, for compositons near the diopside–nepheline and diopside–leucite surfaces in the system diopside–nepheline–kalsilite–silica.

a) Leucitite–leucite trachyte–leucite phonolite–phonolite
b) Leucitite–leucite trachyte–phonolite
c) Leucitite–(nepheline leucitite) leucitophyre–leucite phonolite–phonolite

Trends a and b above are possible for compositions near the plane Di–Ab–Lc, whereas trend c is deduced for those near the Di–Ne–Lc plane.

Washington (1896, 1897) refers to *leucitite–leucite trachyte–leucite phon-*

olite–phonolite associations from Viterbo and Monte Vico regions of Italy. He states (1896b, p. 846) that phonolites, although minor, " . . . occupy in fact an exceptional position occurring only as blocks in the last tuffs ejected by Monte Vico." Washington (1897) proposed that the leucite trend and the associated trachydolerites representing earlier and later differentiates, respectively, were derived from a single potash and lime-rich parental magma.*

These observations lend considerable support to the deduced trends (a and b above) based on experimental studies, and effectively demonstrate the close correspondence between the crystallization behavior of synthetic and natural alkaline melts.

Another *leucitite–leucite phonolite* association is referred to by Holmes and Harwood (1937) from the Sabatinian district of the Roman Province. However, leucite tephrite is a constant member here, which may indicate a modification in the normal leucitic trend due to limestone assimilation (Rittman, 1933) producing anorthite-rich plagioclase in these rocks.

Although possible from theoretical considerations, the association *leucitite–leucitophyre (or simplified nepheline leucitite)–leucite phonolite–phonolite* is not common in nature. Holmes and Harwood (1937) mention an olivine-bearing leucite trend from the Birunga field of Uganda that approximates the above trend, and is referred to in Section 4.2.

The oversaturated sequence from leucitite to trachyte–rhyolite, represents another extreme of fractionation. Such a trend is not well represented in nature.

Peralkaline Trend

Due to the operation of the "plagioclase effect" (Bowen, 1945), the Al content of pyroxene (as Ca- and Mg-Tschermak's or jadeite molecules) and the anorthite content of ternary feldspars may bring about important modifications in the residual liquid trends. Therefore liquids originating close to the Di–Fels divariant surface in the tetrahedron Di–Ne–Ks–Sil (Fig. 26) may, on fractionation, produce another alumina-deficient, alkali-rich residual liquid trend with probable bearing on the comenditic and pantelleritic rocks related to silica-undersaturated parents [the peralkaline trend of Shand (1943)].

3.5 The Leucite–Liquid Reaction and the Pseudoleucite Formation

In certain alkaline rocks leucite is completely replaced by an intergrowth of K-rich feldspar and nepheline or is surrounded by a rim consisting of such an

---

*Magma composition: SiO_2 = 57–58, Al_2O_3 = 17–18 total iron oxides as FeO = 6–7, MgO = 2–3, CaO = 5–6.5, Na_2O = 2–2.5, K_2O = 7–8, H_2O = 1–1.5%.

assemblage. These intergrowths, some of which show zonal arrangements, are described as pseudoleucites. Pseudoleucites normally occur in volcanic rocks but are also found in plutonic rocks, and their formation has been variously explained. Knight (1906) considered that pseudoleucites result from unmixing of Na-rich leucites upon cooling, whereas Bowen (1928) proposed their formation by reaction of early formed leucites with magmatic liquid. The existence of a Lc–Liq reaction point in the system Ne–Ks–Sil, both at 1 atm (Schairer and Bowen, 1935) and at 1 kb /P_{H_2O} (Hamilton and Mackenzie, 1965), supports Bowen's hypothesis.

Fudali (1963), from the consideration of solid solution along the join $NaAlSi_2O_6$–Lc, suggested that the subsolidus breakdown of Na-rich leucites provides a mechanism for the formation of pseudoleucites. The studies of Seki and Kennedy (1964) and Scarfe et al. (1966), on the stability of leucite at elevated pressures, indicate the possibility of the reaction leucite → orthoclase + kalsilite analogous to that proposed by Fudali. However, Seki and Kennedy (1964) stress that water must play an important role in the genesis of pseudoleucites.

The data on the system Di–Ne–Ks–Sil, presented earlier, suggest that the leucite–liquid reaction as proposed by Bowen (1928) is not affected by the presence of diopside and is a possible mechanism for the formation of pseudoleucites in, at least, volcanic rocks. Under volcanic conditions the leucite in equilibrium with liquid will be highly potassic due to the high temperatures involved. However, the possibility of subliquidus breakdown of Na-rich leucite to form plutonic pseudoleucites cannot be discounted, though it requires Na-rich magma as a necessary element (Fudali, 1963).

4 THE SYSTEM DIOPSIDE–FORSTERITE–NEPHELINE–ALBITE–LEUCITE

The leucite and nepheline trends described earlier illustrate the similarities in the crystallization behavior of natural and synthetic residual alkali liquids. It will be appropriate to consider how these trends are related to the mafic magmas of which they are the supposed derivatives. In order to attempt an interrelation of this type, it is essential to examine the relations in the system Di–Fo–Ne–Ks–Sil, forming a portion of the system $CaO–MgO–Na_2O–K_2O–Al_2O_3–SiO_2$.

As the present discussion is mainly concerned with undersaturated alkaline rocks, it is worthwhile to consider only the critically silica-undersaturated portion, Di–Fo–Ne–Ab–Lc, of the system Di–Fo–Ne–Ks–Sil. Di–Ne–Fo–Ab–Lc is simply a leucitic (potassic) extension of the critically undersaturated part of "simple basalt system" of Yoder and Tilley (1962). Such a system encloses all undersaturated compositions from olivine leucitite (ugandite)–leucite basanite–leucite trachytes–leucite phonolite on one hand, to alkali olivine basalts–nepheline basanite–nepheline tephrite–phonolite, on the other. Therefore this system is important in establishing broad correlations between mafic and salic alkaline rocks of diverse compositions.

4.1 Flow Sheet with Univariant and Invariant Relations

The univariant and invariant relations for the system have been reported by Sood and Edgar (1972) from the synthesis of available data and theoretical considerations. The flow sheet is reproduced here in Fig. 30. The necessary information for the construction of this diagram is given in Tables XII and XIII. The arrows indicate the direction of falling temperatures and also define the maximum and minimum on the univariant lines. All the solid phases shown along the univariant lines are complex solid solutions. In Fig. 31 the assemblages given on Fig. 30 are translated into rock names. Figures 30 and 31 in a simple manner summarize the interrelationships of phonolites with olivine leucitites and alkali basalts defining potassic and sodic lineage, respectively.

Some aspects of the flow sheets (Figs. 30 and 31) are noted below:

1. The four quaternary invariant points P, M, R', and S are peritectics.
2. R' is a Lc–Liq reaction point, whereas M, S, and P are Fo–Liq reaction points.
3. The univariant and invariant relations around points R' and S typify sodic trend, whereas that around M and P illustrate a potassic trend.
4. The lowest temperature is that of the minimum M', and there is no quaternary eutectic.
5. The major goal of the liquid compositions in this system is to reach, upon fractionation, the minimum M' following the various univariant lines and the invariant points.
6. The planes Di–Fo–Lc* and Di–Fo–Ab are two thermal divides in the system,

---

*Gupta (1972) has reported a piercing point at $Fo_3Di_{60}Lc_{37}$ and 1296°C. If very small amount of solid solution among the phases is ignored, the system depicts the characteristics of a ternary system.

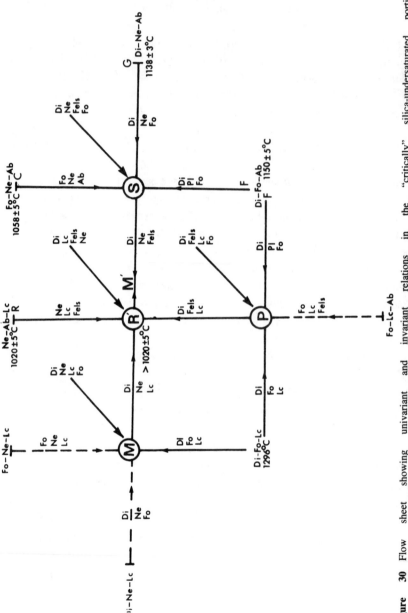

Figure 30 Flow sheet showing univariant and invariant relations in the "critically" silica-undersaturated portion diopside–forsterite–nepheline–albite–leucite of the system diopside–forsterite–nepheline–kalsilite–silica to show the relationships between mafic and felsic rocks. Arrows indicate the direction of falling temperatures. The three solid phases in equilibrium with liquid are listed along the univariant lines. The temperatures of the piercing points and the systems in which the univariant lines originate are shown in parentheses. The join Di–Fo–Lc is close to a compositional plane. (Modified from Sood and Edgar, 1972. With permission of the International Geological Congress.)

Table XII Piercing Points of the Univariant Lines in the System Di–Fo–Ne–Ab–Lc[a]

Number Known	System	Point	Three Solid Phases	Temp. (°C)	References
1	Di–Ab–Lc (Fig. 25b)	P	Di + Lc + Fels	1040 ± 10	Sood et al., 1970
2	Di–Ne–Ab (Fig. 14b)	H	Di + Ne + Pl	1118 ± 3	Schairer and Yoder, 1960a
3	Di–Ne–Ab (Fig. 14b)	G	Di + Fo + Ne	1138 ± 3	Schairer and Yoder, 1960a
4	Ne–Ab–Lc (Fig. 22)	R	Ne + Fels + Lc	1020 ± 5	Schairer, 1950
5	Di–Ne–Fo (Fig. 14c)	A	Fo + Ne + Sp	1245 ± 5	Schairer and Yoder, 1960b
6[b]	Di–Ne–Fo (Fig. 14c)	B	Ne + Sp + Cg	1243 ± 5	Schairer and Yoder, 1960b
7	Fo–Ne–Ab (Fig. 14a)	C	Fo + Ne + Ab	1058 ± 5	Schairer and Yoder, 1961
8	Fo–Ne–Ab (Fig. 14a)	D	Fo + Ne + Sp	1155 ± 5	Schairer and Yoder, 1961
9[b]	Fo–Ne–Ab (Fig. 14a)	F	Ne + Cg + Sp	1290 ± 5	Schairer and Yoder, 1961
10	Fo–Di–Ab (Fig. 10a)	E	Di + Fo + Pl	1150 ± 5	Schairer and Morimoto, 1959
11	Di–Fo–Lc	U	Di + Fo + Lc	1296 ± 5	Gupta, 1972
12	Di–Ne–Lc	—	Fo + Ne + Lc	—	Gupta and Lidiak, 1973
13	Di–Ne–Lc	—	Di + Fo + Lc	—	Gupta and Lidiak, 1973
Possible[c]					
14	Fo–Lc–Ab	—	Fo + Lc + Alk–Fels	—	—
15	Fo–Ne–L	—	Fo + Ne + Lc	—	—

[a]From Sood and Edgar (1972).
[b]Not important in natural systems.
[c]Possible piercing points but no data available yet.

Table XIII Univariant Lines and Invariant Points in the Various Portions of the System Di–Fo–Ne–Ab–Lc

Portions	Possible Univariant Lines	Quaternary	Invariant Points (Figs. 30 and 31)
Di–Ne–Fo–Lc	1. Di + Fo + Lc + Liq 2. Di + Ne + Fo + Liq 3. Fo + Ne + Lc + Liq 4. Di + Ne + Lc + Liq 5. Di + Fo + Lc + Liq	M	Di + Ne + Fo + Lc + Liq
Di–Fo–Ab–Lc	6. Fo + Lc + Alk - Fels + Liq 7. Di + Fo + Pl + Liq 8. Di + Fels + Lc + Liq 9. Di + Ne + Fo + Liq	P	Di + Fo + Lc + Fels + Liq
Di–Ne–Fo–Ab	10. Fo + Ne + Ab + Liq 11. Di + Fo + Pl + Liq 12. Di + Ne + Fels + Liq 13. Di + Fels + Lc + Liq	S	Di + Ne + Pl + Fo + Liq
Di–Ne–Ab–Lc	14. Di + Ne + Fels + Liq 15. Ne + Lc + Fels + Liq 16. Di + Ne + Lc + Liq	R'	Di + Lc + Fels + Ne + liq

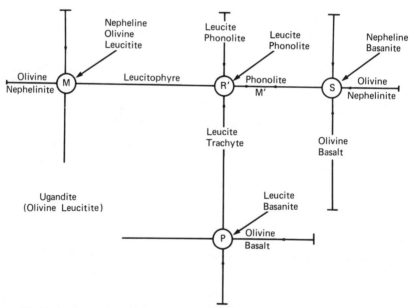

Figure 31 Rock names corresponding to assemblages of Fig. 30. (After Sood and Edgar, 1972. With permission of the International Geological Congress.)

The System Diopside-Forsterite-Nepheline-Albite-Leucite

and the role of Di–Fo–Ab has already been discussed. The system Di–Fo–Lc is considered to represent a "compositional plane," thus separating the highly silica-undersaturated (essentially feldspar-free olivine nephelinites, olivine and nepheline leucitites, etc.) from slightly silica-saturated (feldspar-bearing leucite basanites, leucite trachytes, alkali basalts, etc.) compositions in the system Di–Fo–Ne–Ab–Lc.

4.2 Liquid Trends

Sood and Edgar (1972) have suggested that a parental magma of olivine leucitite (ugandite) composition may be generated on or near the Di–Fo–Lc join. Such a magma can take a more important and common fractionation course to leucite basanite (point P in Figs. 30 and 31). After dissolution of forsterite at P, follow the univariant line PR' giving a leucite-trachyte assemblage. At R' leucite is resorbed and the liquid finally crystallizes at M' to a phonolitic assemblage.

The association *leucitite–leucite basanite–leucite trachyte–leucite phonolite–phonolite* described by Washington (1896, 1897) is a characteristic natural example of the above course. Such a course will be facilitated by a high degree of fractional crystallization.

The olivine leucitite magma may follow another, but less common, fractionation course to *olivine-nepheline leucitites* (point M in Figs. 30 and 31). After complete resorption of forsterite at M, the liquid may follow the univariant line MR' to R', with the liquid finally being consumed to form a phonolite assemblage at M'. This association *olivine leucitite (ugandite)–olivine-nepheline leucitite–nepheline leucitite–leucite phonolite–phonolite*, though rare, is in part described by Holmes and Harwood (1932) from the Birunga field of Uganda (also see Sahama, 1960, 1962, 1974, 1976; Holmes, 1942, 1950, 1952).

A parental magma of the type alkali-olivine basalt or olivine nephelinite can form on or close to the joins Di–Fo–Ab and Di–Fo–Ne, respectively, and may adopt a fractionation course to a nepheline basanite assemblage (point S in Figs. 30 and 31) as described by Schairer and Yoder (1964). At S, after complete resorption of forsterite, the liquid may proceed to M', finally crystallizing to a phonolite. The association *basanite–nepheline tephrite–phonolite* from the Adolf-Fredrich area southwest of Birunga field in Uganda (Holmes and Harwood, 1932) and *olivine nephelinite–nepheline basanite–phonolite* from the Bohemian Mittelgebirge (Knorr, 1932, Scheumann, 1913) are the possible natural analogs. (Also see Bailey, 1974a,b; Sorensen, 1974; Brotzu et al., 1973.)

It is important to note that the consideration of univariant and invariant relations suggest the probable liquid trends with both sodic and potassic affinities, which can result from fractionation of an alkaline melt. The Fo–Liq reaction in the early stages and Lc–Liq reaction in the later stages may, however, exert a dominant control in such liquid lines of descent. The phase relations afford a mechanism of interrelating salic and mafic potassic rocks.

4.3 Generation of Alkaline Magmas

O'Hara (1965, 1968) and O'Hara and Yoder (1963, 1967) suggest a hypothesis for the generation of mafic alkaline magmas by fractional crystallization and melting in the mantle. Their hypothesis is schematically given in Fig. 32.

According to these authors, fractional crystallization at high pressure (ca. 30 kb) of the liquid produced by partial melting of garnet peridotite is believed to give rise to bimineralic (garnet and omphacite) eclogite cumulates and a series

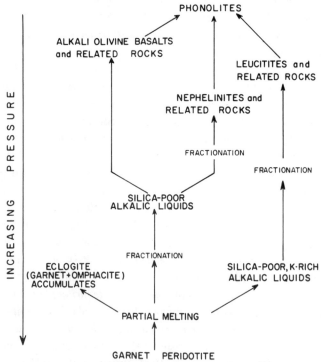

Figure 32 Schematic representation of the model for generation of alkaline magmas. (After Sood and Edgar, 1972. With permission of the International Geological Congress.)

of silica-poor alkalic residual liquids, having the geochemical characteristics of potassic mafic lavas [low SiO_2, usually less than 57%, K_2O/Na_2O ratio greater than 1 (Washington, 1906) high TiO_2 (Borley, 1967; Bell and Powell, 1969; Higazy, 1954)].

Fractional crystallization of the liquid produced by partial melting, under somewhat lower pressures, of the same parent gives rise to silica-poor liquids, which in turn may fractionate to nephelinites and related rocks at low pressures. Under still lower pressure conditions, the same primary parent yields a submagma producing alkali-olivine basalts and related alkaline rock suites.

This hypothesis explains the generation of parental potassic (or leucitic) and sodic (or nephelinic) mafic magmas from a single primary composition in the mantle. Once such magmas are generated, they may follow the crystallization scheme outlined in Figs. 30 and 31 and thus explain the interrelationships of mafic and salic alkaline rocks by processes of crystal–liquid equilibrium and fractionation.

5 THE SYSTEM DIOPSIDE–FORSTERITE–AKERMANITE–LEUCITE

As many potassic rocks contain melilites, Gupta (1972) added akermanite to the system Di–Fo–Lc. The system Di–Fo–Lc–Ak is pertinent to the highly silica-undersaturated melilite-bearing lavas (see Sahama, 1962, 1974, and Velde and Yoder, 1976).

The univariant and invariant relations, on the basis of phase equilibrium studies in the limiting systems, are shown in Fig. 33a,b. Gupta (1972) suggested the system could be considered a "quaternary," due to the small amounts of solid solution in the phases. He concludes that melilite-bearing leucitic rocks could be derived from magmas of olivine-melilitite compositions. However, the K_2O/Na_2O ratio will be the dominant control in producing potassic or sodic lineages.

6 SELECTED SALIC SILICATE SYSTEMS WITH IRON OXIDES

Thus far, in the systems related to alkaline igneous rocks, iron oxides or silicates have not been included as components. However, iron is an important constituent of the pyroxenes and amphiboles, particularly those of peralkaline rocks. Subsequently, phase relations in selected systems related to alkaline rocks will be

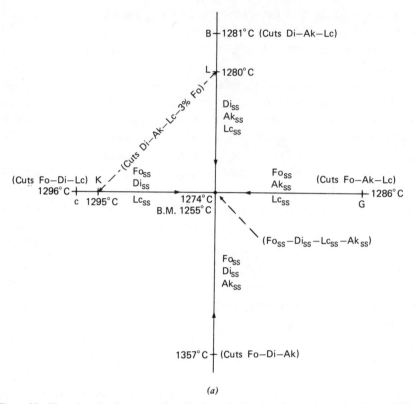

Figure 33 Flow sheet for the system diopside–forsterite–leucite–akermanite. (After Gupta, 1972. With permission of the *American Mineralogist*.) (a) Univariant and invariant relations and phase assemblages.

discussed to evaluate the role of iron and P_{O_2} in the crystallization of alkali melts.

6.1 The System Na_2O–Fe_2O_3–Al_2O_3–SiO_2

Bailey and Schairer (1966) studied the joins:

1 Ac–Jd–$Na_2O \cdot SiO_2$
2 Ac–Ne–Sil
3 Ac–Ne–Ns
4 Ac–Ab–Ns

Selected Salic Silicate Systems with Iron Oxides

5 Ac–Ne–5Na$_2$O·Fe$_2$O$_3$·8SiO$_2$

6 Ac–Ne–Na$_2$O·4SiO$_2$

to interpret the phase relations in the quaternary system Na$_2$O–Fe$_2$O$_3$–Al$_2$O$_3$–SiO$_2$. All the joins have incongruently melting acmite, as a common component. The positioning of the joins covers the major mineral phases found in silica-undersaturated and -oversaturated alkaline rocks with peralkaline affinities. They even suggested reference to the system as the "peralkaline residua system."

Bailey and Schairer (1966) used a Jäneche-type projection (see Ricci, 1966) from the acmite apex to the Na$_2$O–Al$_2$O$_3$–SiO$_2$ base. Because acmite is a component to all the joins, the projections of joins are lines and those of the volumes are areas, thus facilitating the location of volumes and the quaternary invariant

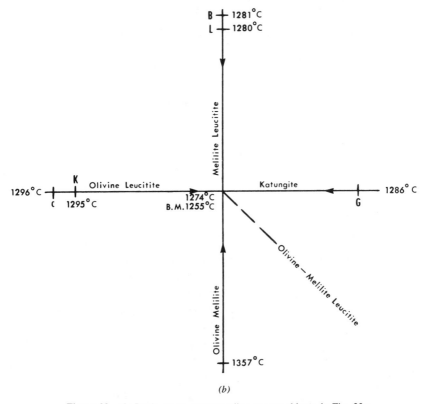

(b)

Figure 33 (b) Rock names corresponding to assemblages in Fig. 33a.

points contained in them. The phase relations data in the six joins were used by Bailey and Schairer (1966) to construct a schematic flow sheet to summarize the univariant and invariant relations, the temperature maxima and minima points, and the compatible and incompatible mineral phases. The flow sheet is reproduced in Fig. 34. Some important observations are:

1. The top part of the diagram shows high temperature relations and the bottom part, the low temperature features. The right-hand side corresponds to silica-oversaturated and the left-hand side to silica-undersaturated compositions.
2. The petrologically important invariant points are the peritectics D and E characterizing Hem–Liq reaction and the eutectics A and B. These invariant points have phase assemblages with close natural analogs.
3. The eutectics A and B correspond to silica-undersaturated and silica-oversaturated parts, respectively.
4. The join Ac–Ab–Ns is the thermal divide between the eutectics A and B, thus separating silica-undersaturated from silica-oversaturated compositions.

The features of the important invariant points, crystallization tendencies of liquids at or near them, and possible petrologic implications may now be summarized. The thermal and related aspects of the invariant points D, B, E, and A are given in Table XIV.

Equilibrium crystallization paths of liquids in the silica-oversaturated portion of the system will be toward the eutectic B (Fig. 34), either directly or through the peritectic D. Liquids on or close to the join Ac–Ab–Sil, but toward the Fe_2O_3 side, will terminate equilibrium crystallization at D. However, fractional crystallization, removal of hematite or its failure to react, will lead the liquids to the eutectic B. The composition of the eutectic B is located toward the silica-undersaturated side of the join Ne–Ac–$Na_2O \cdot 4SiO_2$. Bailey and Schairer (1966) imply that such a composition, upon reaction with aluminous material, will tend to become silica-undersaturated enough to crystallize nepheline. Thus nepheline may crystallize at some granitic contacts independent of desilication of magma by limestone. The quartz trachyte-rhyolite and nepheline trachyte-phonolite associations at the mid-Atlantic ridge may be expressions of such relations.

In the silica-undersaturated portion Ne–Ac–Ab–Ns, liquids, depending on the initial composition and extent of Hem-Liq reaction, approach the eutectic A either directly or through the reaction point E. The composition of the eutectic A is similar to many peralkaline silica-undersaturated rocks. The crystallization

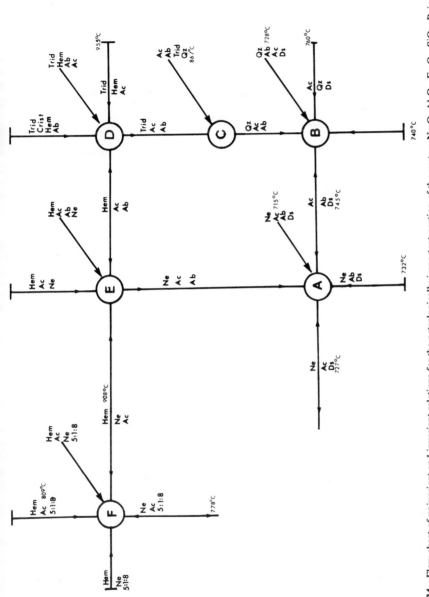

Figure 34 Flow sheet of univariant and invariant relations for the petrologically important portion of the system $Na_2O-Al_2O_3-Fe_2O_3-SiO_2$. Points E and A are quaternary peritectic and eutectic, respectively, for the silica-undersaturated compositions. Point E is also referred to as the Ijolite Point. (After Bailey and Schairer, 1966. With permission of the *Journal of Petrology*.)

Table XIV Characteristic of Invariant Points in a Portion of the System $Na_2O-Fe_2O_3-Al_2O_3-SiO_2$

Invariant Point	Type of Invariant Point	Solid Phases in Equilibrium with Liquid	Temp. (°C)	Location
D	Peritectic for silica-oversaturated compositions	Ac + Hem + Trid + Ab	867–955	Volume bounded by joins A, D, and B
B	Eutectic for silica-oversaturated compositions	Ab + Ac + Ns + Qtz	728	Volume between joins D and F
E	Peritectic for silica-undersaturated compositions	Ne + Ac + Ab + Hem	~908	Volume between joins A and F
A	Eutectic for silica-undersaturated compositions	Ne + Ac + Ab + Ns	715	Volume bounded by joins C, D, and F

of liquids close to the Ac–Ab compositional line has the potential of yielding a derivative liquid corresponding to point E (an ijolite). Bailey and Schairer (1966) have referred to this as the "ijolite point." Therefore very little change in the composition, at or close to E (see Fig. 34), may dictate that liquids move to the quaternary eutectic F (Ne + Ac + Hem + $5Na_2O \cdot Fe_2O_3 \cdot 8SiO_2$) or, with increasing sodium disilicate, to a nepheline syenite type at A (Ne + Ac + Ab + Ns). So liquids near F may give ijolite → foyaite trend at A or ijolite → melteigite trend at F. This may thus provide linkages of ijolite–urtites–melteigites–jacuparingites in a magmatic evolution of alkaline and carbonatite complexes.

The Fe^{3+} substitution for Al^{3+} in Na-feldspar may be a plausible mechanism to cause a change from a silica-oversaturated → silica-undersaturated trend, thus rendering ineffective the thermal divide nature of the alkali feldspar join. Syenites may, therefore, fractionate from silica-oversaturated to -undersaturated rock series through $Fe^{3+} \rightleftharpoons Al^{3+}$ substitutions.

7 LIQUID IMMISCIBILITY IN SILICATE–IRON OXIDE SYSTEMS

Liquid immiscibility is the phenomenon of separation of initially homogeneous liquids into two compositionally distinct liquid phases at a certain stage during or before the crystallization process. The process of "liquid fractionation," however, involves vertical diffusion as a function of pressure and temperature (Hamilton, 1965; Wilshire, 1967).

The phase diagrams of many systems containing silica and one or more metal oxides show the presence of a two-liquid or liquid immiscibility field at 1 atm (see Greig, 1927; Bowen, 1928; Levin et al., 1964; Fischer, 1950; Roedder, 1951, 1956; Schairer and Bowen, 1938; Schairer and Yoder, 1961). Large immiscibility between silicate, oxide, and sulfide liquids has also been observed in many experimental systems (see Naldrett, 1969; McDonald, 1965; Kullerud, 1967) and proposed for natural sulfide–silicate occurrence (Sudbury, Ontario; Insizwa, Africa; Skaergaard, Greenland) through sulfide–silicate liquation process at the early stages of magma formation (Hawley, 1962; Scholtz, 1936; Smith, 1961; Wager et al., 1957).

The petrogenetic importance of liquid immiscibility has been variously proposed on the basis of field and petrographic observations (Loweinson-Lessing, 1955; De, 1974; Fenner, 1948; Holgate, 1954; Marshall, 1914; Ferguson and Currie, 1971, 1972; Tomkief, 1952; Sorensen, 1962, 1969; Drever, 1960), textural and chemical data (Philpotts, 1968, 1970, 1971; Philpotts and Hodgson, 1968; Moore and Calk, 1971; Sood and Ellis, 1973), and mixing of magmas of extreme chemical contrasts (Walker and Skelhorn, 1966; Geijer, 1931; Asklund, 1949; Roedder and Coombs, 1967; Bilibin, 1939).

Kostervangroos and Wyllie (1966, 1968, 1973) demonstrated experimentally the liquation of carbonatitic and mafic alkaline liquids. The recent reports by Roedder and Wieblen (1970, 1971, 1972) on the presence of immiscibility in lunar materials and by Yoder (1973a,c), McBirney and Nakamura (1974), De (1974), Philpotts (1976), and others on the terrestrial materials are significant in reconsidering immiscibility in silicate melts as a viable process (cf. Greig, 1927, and Bowen, 1928). This has stimulated a great interest in further detailed investigations on assessing the extent of liquid immiscibility in silicate melts under different pressure–temperature conditions, as discussed later.

Immiscibility may be due to the chemical or polymerization characteristics of the melt (Hess, 1971). Kogarko and Rhyabchikov (1961) and Kogarko (1964, 1974), on the basis of studies in silicate–halide systems, have suggested that the dissolution of volatiles like F, Cl, S, and P in the melts expand the two-liquid field. This may be related to the substitution of O^{2-} by F^-, Cl^-, PO_4^{3-}, CO_3^{2-}, and SO_4^{2-}, and so on. Thus magmas rich in volatiles may be prone to the immiscibility process.

In discussing theoretical aspects of liquid immiscibility Irvine (1975) has suggested that there is a similarity among two-liquid curves and chemical variation curves, which may be useful in calculating the extent of liquid immiscibility. Irvine (1976) further elaborated from studies in the system Fo–Fa–An–Or–Sil that salic contamination and metastable liquid immiscibility may be important in magmas of intermediate compositions. The nonideal nature of immiscible liquids is worth noting, and the possible formation of silica-rich liquids as immiscible fractions in layered intrusions needs further investigation and confirmation.

Naslund (1976) has reported that liquid immiscibility, in compositions similar to salic magmas, in the system Or–Ab–FeO–Fe_2O_3–Sil increases with increasing oxygen fugacity. The immiscibility is well developed along the univariant curves defining the assemblages Or + Mt + Trid and Mt + Hem + Trid.

Naslund (1977) extended the studies into the system K_2O–CaO–FeO–Fe_2O_3–Al_2O_3–SiO_2 to observe the effect of anorthite. He found that CaO supresses immiscibility better than Al_2O_3. In this system the volume of two liquids rises into the field of anorthite.

Watson and Naslund (1977) from their studies in the system K_2O–Al_2O_3–FeO–SiO_2–CO_2 found that the tendency toward metastable immiscibility is increased with pressure (depth). However, crystallization temperatures also increase with pressure; thus the extent of stable immiscibility must be investigated further. CO_2 is probably more effective in immiscibility than H_2O (see Kogarko in Sorensen, 1974, and Eggler, 1974). Nakamura (1974) observed that immiscibility in the system Or–Fa–Sil disappears at 15 kb.

From their studies Watson and Naslund (1977) imply that pressure and temperature may be similar variables in respect to their influence on melt structure. The variation or change in melt composition in response to decrease in temperature or increase in pressure may be directly related to the polymerization state of structural units of silicates.

The current knowledge of alkaline petrological complexes and studies of alkali-silicate-iron oxide system suggests that liquid immiscibility is a probable petrogenetic process in selected magma compositions and may be particularly effective in alkali magmas and the alkali trends.

CHAPTER 4
Silicate Systems with Volatiles

So far the discussion has centered on describing the phase-equilibrium relations in selected, but petrologically important, silicate systems under anhydrous. 1 atm pressure conditions. Such studies are pertinent to understand and explain the crystallization behavior of magma at or near surface conditions. It is not an oversimplification to assume that the role of volatiles under near-surface conditions does not excessively influence the crystallization tendencies of natural melts. Therefore the deductions and conclusions thus far established are generally valid for low pressure magmatic crystallization.

However, the presence of volatiles, such as CO_2, HF, HCl, H_2S, SO_2, NH_3, and H_2O, in magmas is widely reported in literature, and their importance in magmatic crystallization, at depth, is also greatly emphasized (see Yoder, 1958; Barth, 1962; Oxburgh, 1964; MacDonald, 1972; Roedder, 1965; Green, 1972; Dawson and Powell, 1969; Sigvaldsson and Ellison, 1968; Hill and Boetcher, 1970).

1 EFFECT OF VOLATILES

The collective effect of all the volatiles on magmatic processes is not yet properly quantified, but it seems that it alters the physical state and properties of the magma. Water, undoubtedly, is the most abundant of all the volatiles (see Table I). The effects of water are better known than other volatiles. Water has a lower molecular weight (mol. wt.) than other oxide components of the magma. Because of its low molecular weight, small concentrations of H_2O in natural melts can cause appreciable changes in their physical properties. Goranson (1931) showed that as much as 8 wt. % water could dissolve in granitic melt at 900°C. The solubility of water in silicate melts increases with pressure and feldspathic content (see Chapter 5). A few of the important effects of water are summarized below:

1. Water has a tremendous effect on the lowering of liquidus and solidus temperatures of minerals and rocks (Yoder, 1958) (See Figs. 35–44). At water vapor pressure (P_{H_2O}) of 20 km depth, the liquidus temperature of diopside is lowered by 100°C, whereas for albite it is lowered by as much as 400°C. The liquidus temperature is a function of the solubility of water in the melt. HF even further lowers liquidus temperatures for felsic minerals.

2. The presence of water decreases the viscosity of silicate melts because hydroxyl ion (OH) breaks the oxygen bridges and destroys the Si–O network. On eruption, magmas lose volatiles, due to their decreased solubility with decrease in pressure. This may result in producing viscous lavas.

3. Volatiles are an effective medium for transfer of ions along the crystal lattices. This, in turn, accelerates phase changes, reaction rates, or crystal growths. The presence of large crystals in pegmatites is one good example of crystal growth. Moreover, the experimental runs containing silica-rich compositions reach equilibrium, with little difficulty, in the presence of small amounts of water or other volatiles.

4. The presence of water, at depth, may favor the stability and crystallization of hydrous minerals. Yoder and Tilley (1962) demonstrated from their experimental studies of basalts that at high water vapor pressure, amphibole replaces olivine or pyroxene as the stable liquidus phase. (See also Chapter 6, section 1).

5. Water vapor pressure also decreases the stability limits of certain minerals, for example, leucite is unstable at pressures greater than 2 kb/P_{H_2O}. In other words, orthoclase melts congruently above that pressure.

6. There are changes, with increasing water vapor pressure, in the location and type of invariant points (e.g., Ne–Ks–Sil, Fo–Ne–Sil), univariant lines, and possible fractionation sequences.

The combined effects of H_2O + CO_2 at elevated pressures are just beginning to be determined. The solubility of CO_2 increases with pressure but is related to the water content of the melt. This has implications for the depth of the magma formation. The effects of H_2O + CO_2 may be more important than thus far realized. Liquid immiscibility in silicate melts may also be related to the volatile contents and effective replacement of O^{2-} by other volatile anions.

2 THE SYSTEM DIOPSIDE–ANORTHITE–WATER

Yoder (1965) extended the studies in the system Di–An up to pressures of 5 and 10 kb/P_{H_2O} to determine the effect of water vapor pressure on the binary eutectic in the system.

Figure 35 shows the phase relations in the system at 5 and 10 kb/P_{H_2O}. Also plotted are anhydrous data at 1 atm for comparative purposes. It is obvious from the diagram that the eutectic composition of the system becomes anorthite-rich with increasing water vapor pressure. The eutectic composition under various pressure conditions is given on page 118. (see Fig. 35):

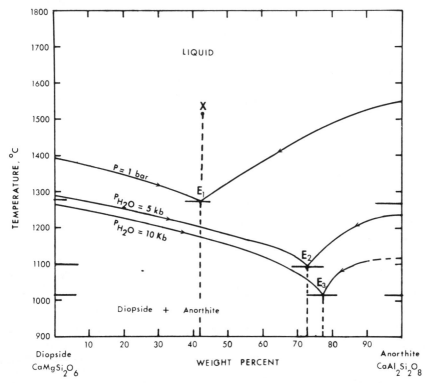

Figure 35 The system diopside–anorthite at various pressures. The temperature of the eutectic is lowered at higher water vapor pressure. The composition of eutectic also shifts toward anorthite with increasing water vapor pressure. An eutectic composition such as X crystallizes Di + An together at 1 bar, whereas diopside is primary phase at elevated water vapor pressure. (5 and 10 kb/ P_{H_2O} data from Yoder, 1965. With permission of the Carnegie Institution of Washington.)

1   $Di_{58}An_{42}$ at 1 bar (anhydrous)
2   $Di_{27}An_{73}$ at 5 kb/P_{H_2O}
3   $Di_{22}An_{78}$ at 10 kb/P_{H_2O}

Petrologically this represents a compositional change from a gabbro to an anorthosite and a possible explanation for the formation of anorthosites or feldspar-rich xenoliths in calc-alkaline rocks as a normal evolution of the magmatic equilibrium process (Yoder, 1969).

Consider a composition such as point X (Fig. 35). The increase in water vapor pressure will change the crystallization path from coprecipitation of Di + An at 1 atm to the separation of diopside alone, over a long temperature interval. If we assume fluctuating water vapor pressure conditions in the magma chamber, then there is a basis to explain an alternating precipitation of diopside (a pyroxene) and anorthite (a plagioclase), or coprecipitation of phases or delayed precipitation of pyroxene or plagioclase phases. This has probable implications to the gabbro-anorthosite associations in layered intrusions. The sudden decrease in pressure due to rapid magma ascent may result in decreased solubility of volatiles in the melt and cause explosive volcanism in specific compositions.

3 THE SYSTEM ALBITE–ANORTHITE–WATER

The system Ab–An–H_2O has been studied by Yoder et al. (1957) to a pressure of 5 kb/P_{H_2O}. The liquidus phase relations at various pressures are shown in Fig. 36.

One important aspect of the phase diagram is the large depression (~400°C) in the liquidus surface in response to increasing water vapor pressure. The looplike arrangement is still present at various water vapor and anhydrous pressure conditions.

It is petrologically interesting to note that for a constant temperature the composition of plagioclase feldspar becomes increasingly anorthitic with increasing water vapor pressure. Consider a constant temperature of 1200°C. The composition of feldspars crystallizing at various pressures is as follows (see Fig. 36):

1   At 1 bar, An_{36} at point P_1
2   At 150 bar/P_{H_2O}, An_{52} at point P_2
3   At 5 kb/P_{H_2O}, An_{95} at point P_3

The System Albite-Anorthite-Water

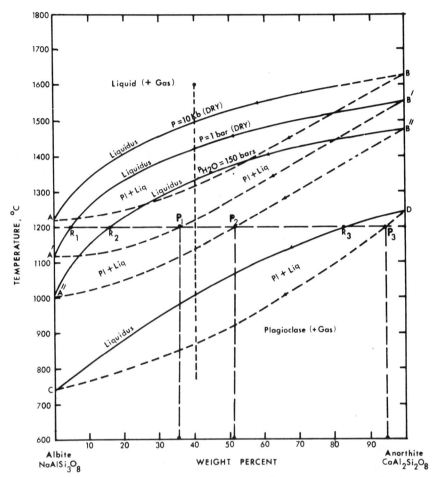

Figure 36 The system albite–anorthite at 1 bar (Bowen, 1913), at 150 bar/P_{H_2O} (after Yoder, 1969), and at 5 kb/P_{H_2O} (after Yoder et al., 1957). (With permission of the Carnegie Institution of Washington.) Note the lowering of liquidus temperature by over 400°C.

This implies that the compositional zoning in plagioclase feldspars may also be related to crystallization under different or fluctuating water vapor pressure conditions. The high calcium cores may represent initiation of crystallization at depth, which progressively (abruptly or alternatingly) become low calcium with crystallization at shallow depth but without substantial changes in the thermal conditions. The occurrence of any abrupt (or alternating) compositional zoning in feldspars may thus reflect fluctuating, nonequilibrium changes prevalent in the magma chamber. Yoder (1969) used the relations in the system as a basis to suggest that the occurrence of high An-plagioclase and the oscillatory zoning in feldspars of the calc-alkaline rocks is probably a reflection of normal nonequilibrium crystallization under varying water vapor pressure conditions rather than other complex mechanisms.

The sudden drop in pressure in the magma chamber or rapid ascent of magma will decrease the solubility of water in the melt and cause explosive volcanism or oxidation of iron.

The anhydrous pressure seems to exert an effect opposite to that of water vapor pressure conditions that can be seen by comparing feldspar compositions at 1400°C (Fig. 36).

4 THE SYSTEM ALBITE–ORTHOCLASE–WATER

The phase relations in the anhydrous system Ab–Or are shown in Fig. 37. It is apparent that alkali feldspars form a continuous solid solution series at high temperature at 1 atm pressure. There is, however, a significant field of leucite, due to the incongruent melting of orthoclase.

Below 700°C the system shows a two-feldspar field called solvus. The solvus has a critical point of unmixing at 660°C and characterizes the area of perthites (Fig. 37). Alkaline rocks containing perthite must then crystallize at low water vapor pressure. Perthite may thus be a good geothermometer in such cases.

The effects of increasing water vapor pressure are also included in Fig. 37. It is seen that the liquidus surface is depressed by almost 400°C. The leucite field is also diminished. Above 2 kb/P_{H_2O} the leucite field disappears, suggesting absence of leucite in plutonic alkaline rocks.

Morse (1968b) has studied the system up to a pressure of 5 kb/P_{H_2O} and the equilibrium diagram is reproduced in Fig. 38. There are large areas of albite and orthoclase solid solution. The solvus and the liquidus surfaces intersect each other, thus inhibiting a continuous solid solution in the alkali feldspars at high pressures. It is implied that two distinct feldspars may crystallize in deep-seated plutonic rocks. Equilibrium crystallization of liquid compositions for this system will be somewhat similar to that described for the system Ab–An in Chapter 2.

A melt of composition X (or Y, Fig. 38) will begin crystallization with crystals of Ab_{98} (or Or_{92}). The composition of crystals will change continuously, by reaction with the liquid, to a somewhat more potassic (or sodic) feldspar. The crystallization will cease when crystals of feldspar of composition as that of the original liquid (X or Y) crystallize.

The System Albite-Orthoclase-Silica-Water

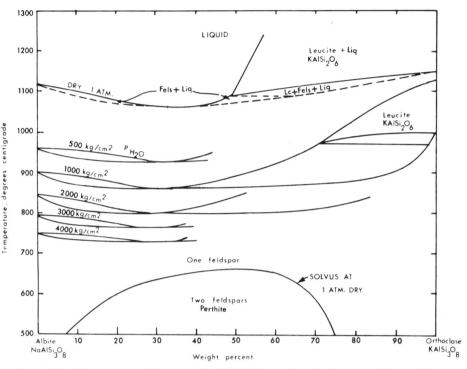

Figure 37 Albite–orthoclase: The system of alkali feldspars at 1 bar (after Bowen and Tuttle, 1950; with permission of the *Journal of Geology*) and to 4 kb/P_{H_2O} (after Tuttle and Bowen, 1958; with permission of the Geological Society of America). The field of leucite diminishes with increasing water vapor pressure and disappears above 2 kb/P_{H_2O}.

However, a liquid of composition E (or those approaching E by fractionation) will begin simultaneous separation of two distinct feldspars of composition B and C (Fig. 38). With falling temperature there will be the tendency for limited exsolution (perthite formation), which can be seen from the shape of the solvus. However, perthites will not form in rocks formed at high water vapor pressure.

5 THE SYSTEM ALBITE–ORTHOCLASE–SILICA–WATER: THE GRANITE SYSTEM

Let us add the silica component to the system of alkali feldspars. Such a system is applicable to the rocks of granite family as it closely approximates most granitic compositions. Therefore the phase relations in the "granite system" at

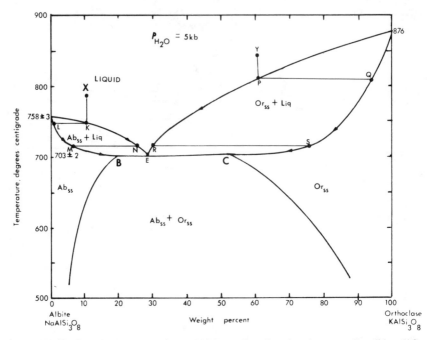

Figure 38 The system albite–orthoclase at 5 kb/P_{H_2O} where the solvus intersects the solidus. (After Morse, 1968. With permission of the Carnegie Institution of Washington.) KLNM and PQRS are crystallization paths for compositions X and Y respectively. Note the lowering of melting temperature of more then 400°C compared to anhydrous conditions.

various water vapor pressure conditions are of prime importance for understanding the evolution of granites and pegmatites.

5.1 General Features and Crystallization Trends

Tuttle and Bowen (1958) have produced a remarkable work on the system Ab–Or–Sil–H_2O up to 4 kb/P_{H_2O}. The liquidus surfaces for the system at 1 atm and various water vapor pressure are presented in Fig. 39. The phase relations point out the following:

1. The field of leucite decreases with increasing water vapor pressure. It completely disappears above 2 kb/P_{H_2O} (see Fig. 39b).
2. The temperature of the liquidus surface is depressed with increasing water vapor pressure and affects the stability of silica polymorphs (Fig. 39b).

The System Albite-Orthoclase-Silica-Water

3 The Alk–Fels–Sil phase boundary is replaced by a ternary eutectic above pressures of 3.5 kb/P_{H_2O}. In other words, the system is "truly" ternary above that pressure (Fig. 39c).

4 The temperature of the eutectic is 625°C (Fig. 39d), similar to the minimum melting curve for granites.

5 The composition of the eutectic corresponds closely to the normative feldspar and quartz contents of granites.

6 The eutectic relations demonstrate that two distinct alkali feldspars can directly crystallize from granitic melts under plutonic conditions.

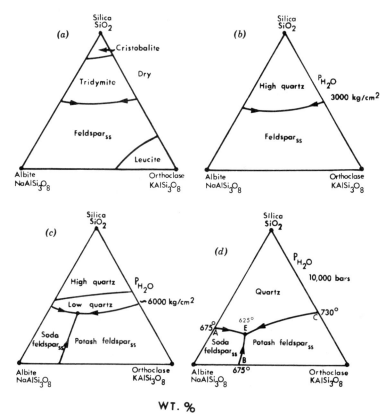

Figure 39 (a) at 1 bar (after Schairer, 1950; with permission of the *Journal of Geology*); (b) at 3 kb/P_{H_2O} (after Tuttle and Bowen, 1958; with permission of the Geological Society of America); (c) at 6 kb/P_{H_2O}; and (d) at 10 kb/P_{H_2O} (after Luth et al., 1964; with permission of the *Journal of Geophysical Research*.) Note the system is ternary at pressures above 5 kb/P_{H_2O}.

7 With increasing water vapor pressure, the minimum melting composition shifts toward the albite apex.

The eutectic relations in the system at elevated water vapor pressures simplify the discussion of crystallization paths for various melts in the system. A liquid composition, depending on the field it plots in Fig. 39d, will begin crystallization with either quartz, Na-feldspar, or K-feldspar phases. With continued cooling, it will approach the eutectic E, following one of the cotectic curves (AB, BE, or CE). At E, Or + Ab + Qtz crystallize simultaneously, as the temperature and composition remain constant and the liquid phase is eliminated. Only liquid compositions lying on the cotectic curves will begin coprecipitation of two solid phases. The proportion of the phases at any stage can be determined by using the Lever Rule. The final proportions of the solids will be related to the initial composition.

On the basis of phase relations in the systems Ab–Or–H_2O, Ab–Or–Sil–H_2O, and the type of feldspar phase present, Tuttle and Bowen (1958) proposed that granites may be classified as *hypersolvus* or *subsolvus*. The hypersolvus granites contain one phase of alkali feldspar and the subsolvus granites have two distinct phases of alkali feldspars.

The equigranular, hypidiomorphic, and graphic textures common in granites, may be a reflection of crystallization of natural granitic melts near the eutectic point (or cotectic boundaries close to this point) in the synthetic system. Most normative compositions of the granites plot close to the eutectic E in the granite system (Fig. 39d) at high water vapor pressure, whereas rhyolite, cluster near the "minimum" at 1 atm pressure (see Fig. 22b). However, Luth et al. (1964) have shown that alkali granites (acmite and sodium silicate normative) do not plot near the eutectic at elevated water vapor pressure. This implies that alkali granites (pantellerites*) may crystallize under water-unsaturated conditions and may be directly derived from "dry" basic magmas (Luth et al., 1964).

The low temperature (625°C) of the eutectic in the system Ab–Or–Sil–H_2O and its compositional similarity with the granites suggest that, in addition to the formation of granites by differentiation of basaltic magmas, granitic melts may also be produced by partial melting of crustal and subcrustal rocks as the temperatures are within the range of the minimum melting curve for granites. There-

---

*For further discussion of peralkaline silicic trends, the reader is referred to the experimental studies of Carmichael and MacKenzie (1963), Bailey and Schairer (1964), and Thompson and MacKenzie (1967).

fore anataxis, palingenesis, or partial melting may be a feasible culmination of ultrametamorphism where field, petrochemical, and petrographic evidences are also supportive or conclusive in this regard. It is further implied that granitic magmas in their pegmatite derivative stages are probably saturated with water at the start of the crystallization process. The coarsely crystalline texture of pegmatites may be an indication of such a physical state.

6 THE SYSTEM ANORTHITE–ALBITE–ORTHOCLASE–WATER: THE SYSTEM OF FELDSPARS

Feldspars are important constituents of most igneous rocks. Their compositions and proportions provide a major basis of classification of igneous rocks. The composition of feldspars in common igneous rocks is shown in Fig. 40.

The phase relations in the systems of alkali feldspars and plagioclase feldspars have been separately discussed in the previous sections. The individual systems show interesting solid solution tendencies as a function of pressure and temperature, with implications as to the probable use of feldspars as geothermometers.

The phase relations in the system An–Ab–Or–H_2O, representing the addition of anorthite component to the system of alkali feldspars, are useful in understanding the collective crystallization behavior of all feldspars and the extent of solid solution they exhibit. As described earlier, the plagioclase feldspars show complete solid solution up to a pressure of 5 kb/P_{H_2O}, whereas the alkali feldspars show complete solid solution only at low pressures and high temperatures, with unmixing at low temperatures. Moreover, the miscibility in alkali feldspars somewhat decreases with increasing water vapor pressure, due to expansion of the solvus and its intersection with the solidus. The system An–Or has limited solid solution under all conditions. The equilibrium liquidus relations for the system An–Ab–Or at 1 atm are shown in Fig. 41a, whereas Fig. 41b shows liquidus surface at 5 kb/P_{H_2O}. The diagrams are based on the studies of Franco and Schairer (1951), Yoder et al., (1957), Tuttle and Bowen (1958), Luth et al. (1964), James and Hamilton (1969) and Morse (1970).

A few observations can be made:

1 The surface of crystallization is substantially lowered with increasing water vapor pressure. Note the temperature of the minima and isotherms at 1 atm and 5 kb/P_{H_2O} (Fig. 41a,b). It is lowered by more than 400°C.
2 The field of leucite contracts and disappears at 5 kb/P_{H_2O} (Fig. 41a,b).

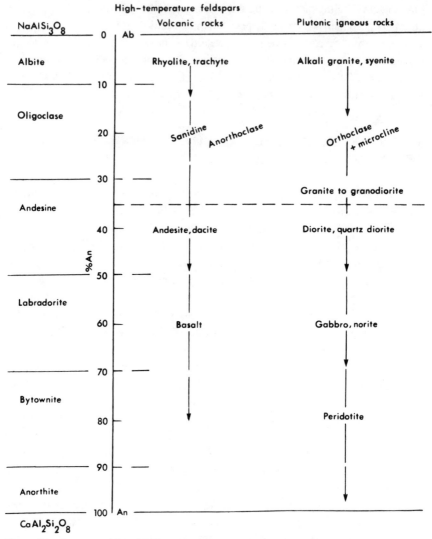

Figure 40 The composition of feldspars in common igneous rocks. (After Hyndman, 1972. With permission of McGraw-Hill Book Co., New York.)

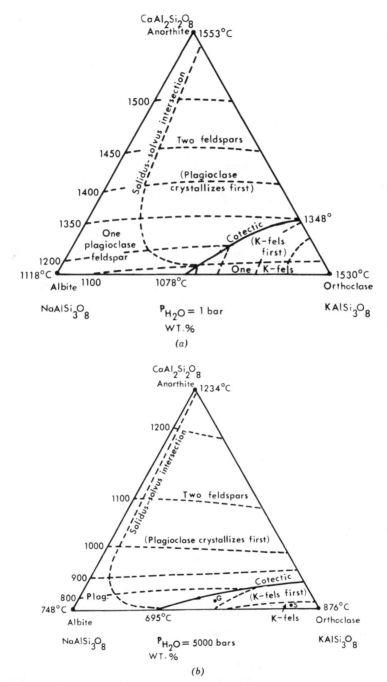

Figure 41 Orthoclase–albite–anorthite: The system of feldspars. *(a)* Phase equilibrium relations at 1 bar/P_{H_2O} (after Franco and Schairer, 1951; with permission of the *Journal of Geology*). Leucite field is omitted. *(b)* Phase equilibrium relations at 5 kb/P_{H_2O} (after Yoder et al., 1957; with permission of the Carnegie Institution of Washington).

3 The solvus–solidus intersection and the limited solid solution among anorthite and orthoclase highlights the areas of two feldspars, one plagioclase feldspar and one K-feldspar.
4 The solubility among feldspar components decreases with increasing water vapor pressure. In other words, the two-feldspar region is enlarged (Fig. 41b).
5 The solubility of anorthite in alkali feldspars decreases with increasing water vapor pressure. According to Hamilton (1969), the solubility of anorthite in alkali feldspars is 5–7% at 900°C, 0.5 kb/P_{H_2O} and 3–5% at 700°C, 1 kb/P_{H_2O}.

Depending on the initial composition, a melt may crystallize two feldspars, a plagioclase feldspar and a K-rich feldspar. The composition, however, will continuously change with further crystallization and crystallization paths will be unique for each composition.* (Also see Hyndman, 1972.)

It may be pointed out that alkaline rocks with K-rich feldspar (and no plagioclase) suggest crystallization from a magma with a composition plotting in the narrow Ca-poor, one K-feldspar field (point S in Fig. 41b). The subsequent slow cooling will undoubtedly produce some perthite lamellae, due to the slope of the solvus (see Fig. 38). However, a felsic rock with K-feldspar phenocrysts and K-feldspar and plagioclase in the matrix will indicate crystallization from a magma whose composition lies in the K-feldspar field (point G in Fig. 41b).

7 THE SYSTEM ANORTHITE–ALBITE–ORTHOCLASE–SILICA–WATER

Alkali feldspars of many granitic rocks tend to contain significant amounts of CaO probably as an anorthite component. Logically, von Platten (1965) and Winkler (1967) extended the studies in the granite system, with the addition of anorthite to see its effect on the melting behavior, particularly the position of the low melting region and its relation to anataxis.

Von Platten (1965) investigated a group of compositional planes within the system An–Ab–Or–Sil–H_2O at a fixed pressure of 2 kb/P_{H_2O}. He studied four

---

*For theoretical and experimental discussion of crystallization paths in three-phase and ternary solid solution systems, the reader is referred to Stewart and Roseboom (1962), MacKenzie and Rahman (1969), Roeder (1974), and Zavaritskii and Sobolev (1966).

The System Anorthite-Albite-Orthoclase-Silica-Water

planes with Ab/An ratios ranging between 1.8 to 7.8. Figure 42 is a projection from the An apex on the Ab–Or–Sil–H$_2$O base and shows the effect of the various anorthite contents on the ternary minimum.

One major observation is that the temperature of the minimum or the beginning of melting increases with increasing anorthite content or decreasing Ab/An ratios. The increase in temperature is 35°C, compared to the studies of Tuttle

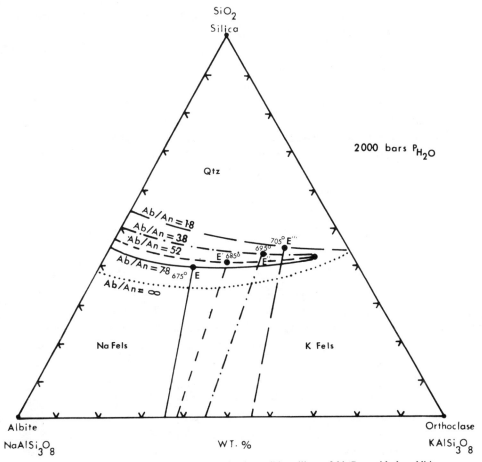

Figure 42 The phase relations in the system orthoclase–albite–silica at 2 kb/P_{H_2O} with the addition of various proportions of anorthite. The composition of the eutectic E shifts toward Or–Sil line with decreasing albite/anorthite ratios whereas its temperature increases by 30°C (see point E', E'', E''' . . .). (After Von Platten, 1965. With permission of Oliver & Boyd, Edinburgh, Scotland.)

and Bowen (1958) in the granite system with no anorthite. The shift in composition of the minimum is toward the silica-orthoclase line.

In granites the feldspars are more calcic than albite. This would explain why the normative quartz-feldspars plots of granites in the Ab–Or–Sil–H_2O system are a bit skewed to the orthoclase side (see Tuttle and Bowen, 1958; Presnall and Bateman, 1973).

8 THE SYSTEM NEPHELINE–KALSILITE–SILICA–WATER: THE SILICA UNDERSATURATED PORTION

The system, as previously described (See Chapter 3, section 2), consists of two portions. The silica-oversaturated portion Ab–Or–Sil–H_2O, above the alkali feldspar join, has been discussed in Sections 5 and 7. The silica-undersaturated portion, below the alkali feldspar join, has been studied at 1 and 5 kb/P_{H_2O} by Hamilton and MacKenzie (1965) and by Morse (1968a), respectively.

The liquidus phase equilibrium relations for the system Ne–Ks–Sil–H_2O at 1 and 5 kb/P_{H_2O} are shown in Figs. 43 and 44, respectively. Some important observations on the phase relations (see Fig. 44) are noted below:

1 The field of leucite contracts at 1 kb/P_{H_2O} and is totally eliminated at 4 kb/P_{H_2O}.
2 The Ab–Or thermal divide becomes more prominent and defined with increasing water vapor pressure.
3 The increasing water vapor pressure somewhat depresses the composition of the minimum M in the silica-undersaturated portion toward the Ne–Ks base (cf. Fig. 22a and 43). The composition of M is not, however, drastically changed. This implies that the feldspathoidal syenites may show only minor compositional variation in the geological environment of their formation. Such is also the case observed in nature (Morse, 1968a).
4 The liquidus surface is substantially depressed with increasing water vapor pressure, as seen from the temperatures of the various minima and the isothermal lines.
5 The reaction point R at 5 kb/P_{H_2O} is characterized by Ab + Liq → Anl + K-Fel reaction rather than Lc + Liq reaction at low pressures (see inset Fig. 44).

The System Nepheline-Kalsilite-Silica-Water

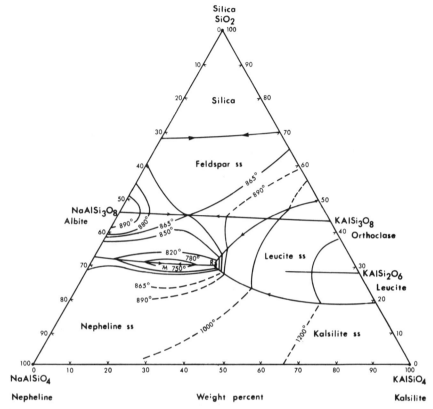

Figure 43 The Petrogeny's Residua system nepheline–kalsilite–silica at 1 kb/P_{H_2O} (after Hamilton and Mackenzie, 1965; with permission of the *Mineralogical Magazine*). Note the decrease in melting temperatures of various compositions and points (cf. Fig. 22a). The field of leucite is also diminished.

6 The fields of nepheline and Na-feldspar are separated by a small field of analcite. Thus Ne + Ab is not a stable assemblage at high water vapor pressure. The rocks containing such an assemblage must then form at pressures lower than 5 kb/P_{H_2O} (Morse, 1968a).

7 At 5 kb/P_{H_2O}, the minimum M is replaced by an eutectic E', where K-Fels + Ne + Anl + Liq is a stable assemblage.

8 The temperature of the minimum is lowered from ~1020°C at 1 atm to 750°C at 1 kb and 635°C at 5 kb/P_{H_2O}.

9 The two-nepheline region seems to be absent at water vapor pressure of 5 kb/P_{H_2O}.

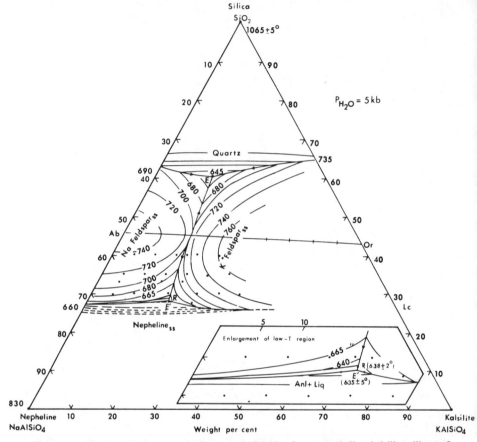

Figure 44 The phase relations in the Petrogeny's Residua System nepheline–kalsilite–silica at 5 kb/P_{H_2O}. (After Morse, 1968. With permission of the Carnegie Institution of Washington.) There are eutectic points in the silica oversaturated and silica undersaturated portions at 645° and 638°C, respectively. Point R shows an analcite + liquid reaction instead of leucite + liquid reaction (see Figs. 22a and 43). Albite–orthoclase thermal divide persists even to 5 kb/P_{H_2O} conditions. Inset shows enlargement of the portion around points E' and R.

At 5 kb/P_{H_2O}, the trends of liquid composition in the silica-undersaturated portion will be to reach, upon crystallization, the eutectic E' (Fig. 44). The liquids in the Na-feldspar field (and those above the line R–Or in the K-feldspar field) will move toward R along the two-feldspar boundary. At R, Ab + Liq reaction produces Anl + K-Fels and the liquid finally terminates crystallization at E' to give Ne + Anl + K-Fels assemblage. The melt in the nepheline field,

The System Nepheline-Kalsilite-Silica-Water

depending on the initial composition, will tend to approach E' by following the nepheline–K-feldspar or nepheline–analcite phase boundaries to give a final assemblage of Ne + Anl + K-Fels.

Morse (1968a), using his and the other published data on the system Ne–Ks–Sil under different water vapor pressure conditions, constructed a profile along the Fels + Liq boundary curve from the granite minimum through feldspar join and to the nepheline syenite or phonolite minimum. Figure 45 shows the projection of profiles, from 1 atm pressure of 10 kb P_{H_2O}, parallel to the lines of equal silica content and Ne/Ks ratio of one.

It may be concluded from the distribution of the profiles that the thermal

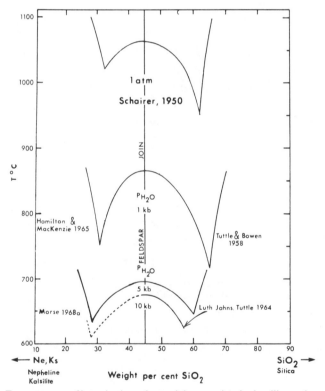

Figure 45 Temperature profile projections along minimum points in the silica undersaturated and oversaturated portions of the system nepheline–kalsilite–silica at various pressures. Note the hump separating the two temperature minima indicating that the albite–orthoclase join is a temperature maxima under all conditions. (After Morse, 1968a. With permission of the Carnegie Institution of Washington.)

divide on the feldspar join is not eliminated in the geologically probable conditions of water vapor pressures for the formation of the felsic rocks. Therefore critical silica undersaturation and oversaturation may be maintained during fractionation, thus resulting in the spatial separation of granites and nepheline syenites. It is unlikely that water undersaturation will raise the temperatures of either granite or nepheline syenite minima above that of the beginning of melting temperatures for syenites (Morse, 1968). Ignoring some offsetting caused by the An contents of the feldspars, the syenitic rocks from a *single province* will generally plot rather close to the thermal divide. MacKenzie (1972) reported that the trachytic rocks from the widely separated Atlantic islands show uneven distribution near the feldspar join, and many with high An content of feldspars plot quite far from it.

Tilley (1957) suggested the presence of iron (or mafic minerals) may render the feldspar thermal divide inoperative. The approach of the pyroxene phase volume very close to the feldspar surface in many synthetic systems (see Edgar, 1964, Edgar and Nolan, 1966; Morse, 1968a, Sood et al., 1970) may be interesting in this regard. Therefore there is at least a tendency by the mafic components toward intersection of the feldspar surface, which may be further enhanced in the multicomponent natural systems. In other words, weakening of the effect of the thermal divide may be possible. The spatial separation of silica-undersaturated and -oversaturated rocks in the petrologic complex must, however, be kept in mind.

The current knowledge of the synthetic silicate systems supports the derivation of syenitic rocks through crystallization–differentiation of basic alkali magmas. The presence of syenites in the deep-seated complexes at Kinglapait, Labrador (Morse, 1969) and Insche, Scotland (Read et al., 1961; Clark and Wadsworth, 1970) are among the relatively small number of areas to clearly demonstrate their formation through the process of fractionation. Morse's (1968a, p. 112) statement, "their small volume and frequent well-defined association with logically parental series render their derivation and fractionation from basalts quite credible," sums up the problem well. In most cases the low pressure basaltic differentiation produces either rhyolitic or phonolitic residues.

9 THE SYSTEM FORSTERITE–ANORTHITE–ALBITE–SILICA–WATER

Kushiro (1974) has studied the system Fo–An–Ab–Sil–H_2O at 15 kb/P_{H_2O} to assess the role of partial melting in a hydrous mantle to form calc-alkaline (andesite) magma. The liquidus phase relations for the portion Fo–Ab$_{50}$An$_{50}$–Sil–H$_2$O of the system Fo–An–Ab–Sil–H$_2$O at 15 kb/P_{H_2O} are shown in Fig. 46. Besides prominent regions of forsterite and enstatite, it has the regions of amphibole, plagioclase, and quartz. It has a piercing point A at 1000 ± 20°C, where Fo + En + An coexist with liquid and vapor. The other piercing point, B, is estimated for the 10% Or plane in this system with the

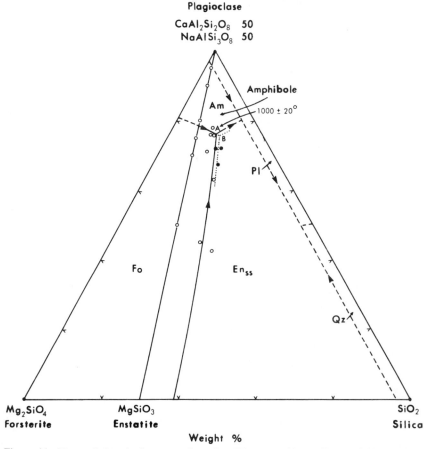

Figure 46 Phase relations in the system forsterite–albite$_{50}$–anorthite$_{50}$–silica at 10 kb/P_{H_2O} with 10 wt. % orthoclase. Point A is at 1000 ± 20°C and has Fo + En$_{ss}$ + Am + Liq + vapor in equilibrium. Point B is the estimated piercing point at 10 wt. % orthoclase plane of the system forsterite–plagioclase–orthoclase–silica at 15 kb/P_{H_2O}. The dashed lines near the plagioclase–silica join are not fully determined. (After Kushiro, 1974. With permission of the Carnegie Institution of Washington.)

addition of orthoclase. The composition of the piercing points is given in Table XV. The temperature of point B is less than 1000°C, and its composition is more silica-rich than A. In both cases the composition is similar to that of andesites. These results support the hypothesis that calc-alkaline magma may be produced by partial melting of hydrous upper mantle (see Yoder, 1969; O'Hara, 1965) at least in the 15 kb depth zone. McBirney (1969) has suggested that the breakdown of hydrous minerals in the subduction zone will release water, which may bring the mantle above within the range of temperature of melting. Such a process could produce calc-alkaline magma. The physicochemical parameters are similar to those obtained by Kushiro (1974) in the system discussed here.

10 OTHER SELECTED SILICATE–WATER–CARBON DIOXIDE SYSTEMS

Considering the importance of iron in the pyroxenes of alkaline rocks, Nolan (1966) and Edgar and Nolan (1966) investigated a series of planes in the system Di–Ac–Ne–Ab up to a pressure of 2 kb/P_{H_2O} and controlled partial pressure of oxygen (P_{O_2}). They report that the pyroxene compositions exert important controls in the positioning of the low melting points. There is also a good agreement between rock compositions and the low melting regions in the planes $Di_{50}AC_{50}$–Ab–Ne, $Di_{20}Ac_{20}$–Ab–Ne, and $Di_{10}Ac_{90}$–Ab–Ne. The plot of agpaitic rocks (Na + K/Al $\geq$ 1) containing aegirine-rich pyroxenes, due to excess of

Table XV Compositions of the Piercing Points in the System Fo–$An_{50}Ab_{50}$–Sil and 10 wt. % Or[a]

	Piercing Point A	Glass	Piercing Point B	Norm of Composition B
SiO_2	59.7	58.6	60.8	7.7 quartz
Al_2O_3	21.3	22.7	22.8	10.0 orthoclase
MgO	6.88	6.53	5.93	33.8 anorthite
CaO	7.66	8.19	6.80	33.8 albite
Na_2O	4.49	3.93	3.99	14.8 enstatite
K_2O	—	—	1.69	
Total	100.03	100.0	100.01	

[a] From Kushiro (1974).

Other Selected Silicate-Water-Carbon Dioxide Systems

Na, corresponds very well to the minimum in the plane Di_5Ac_{95}–Ne–Ab. [Also see Ford (1972) for data on the system Ac–Or–$Na_2O\cdot4SiO_2$–H_2O.]

Yoder and Upton (1971) studied the Di–San system at 5 and 10 kb/P_{H_2O}. They report the occurrence of the cotectic points at Di_2San_{98} and Di_1San_{99} at water vapor pressures of 5 and 10 kb/P_{H_2O}, respectively. They suggest that the assemblage Di + San should be stable up to depth of at least 40 km, which may explain the presence of mafic xenoliths in trachytes. Therefore, it is possible that the parental materials, on partial melting, could produce trachytic (syenitic) magmas. The unique place of trachytes (syenites) in igneous complexes must, however, be considered.

Kushiro (1969a) extended the high pressure studies in the system Fo–Ne–Sil in the presence of water to assess the effect of water on the composition and the temperature of the beginning of melting of the piercing points in the system compared to the confining pressure studies (see Kushiro, 1968). He observed that the composition of the first liquid is pressure-dependent both in dry and wet systems at high pressures. At 17.5 kb/P_{H_2O}, the composition of the first liquid from the melting of Fo + En + Jd(Ab) mixtures is silica-saturated. However, at 20 kb/P_{H_2O} the composition of the first liquid is critically silica-undersaturated (Ne-normative). He further stressed that with continued partial melting, at 20 kb, the composition of the liquid moves to Ol + Hy-normative to Qtz-normative. Orthopyroxene may thus crystallize from Ne-normative liquids above pressures of 20 kb (hydrous or anhydrous). Eggler (1974) found that the addition of CO_2 to the system Fo–Ne–Sil changes the liquidus phase from olivine to orthopyroxene at 22.5 kb instead of 30 kb dry. The liquid composition at the invariant point becomes more silica-undersaturated and very much different than in the presence of water (see Figs. 47 and 59). It is therefore apparent that the magma composition can be related to volatile contents, depth, and degree of melting in the mantle. The addition of diopside or other components* considerably affect the liquid compositions.

Luth (1967) investigated the system Fo–Ks–Sil–H_2O up to pressure of 3 kb/P_{H_2O}. The system contains Fo + Ph + Liq and En + Ph + Liq phase boundary curves. This explains the brown mica rimming of forsterite or pyroxenes found in rocks derived from potash-rich basic magmas. In other words, potassic rocks may be a product of biotite resorption as suggested by Bowen (1928).

Wenlandt (1977a) extended the studies on the system Fo–Ks–Sil to 30 kb

---

*The reader is referred to Kushiro (1969) and Eggler (1974) for the discussion of the systems Fo–$CaAl_2SiO_6$–Sil and Fo–Ne–$CaAl_2SiO_6$–Sil with H_2O + CO_2.

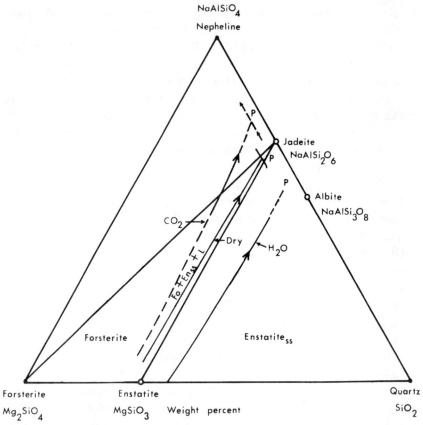

Figure 47 Estimated shifts in the forsterite–enstatite phase boundary in the system forsterite–nepheline–silica at 20 kb for anhydrous, excess water and carbon dioxide conditions. The effect of carbon dioxide is to push the boundary toward silica undersaturation and that of water to silica saturation. (After Eggler, 1974. With permission of the Carnegie Institution of Washington.)

with the addition of CO_2, using the data of Eggler (1975) and Lindsley (1966) to assess the generation of alkaline magmas by partial melting of the mantle periodotite. He found that the field of enstatite expands with increasing pressure well beyond the critical silica undersaturation join Fo–San. Near 14 kb for CO_2-saturated conditions, the liquid compositions in equilibrium with Fo + San + En are leucite normative. The increase in pressure makes the liquids increasingly alkaline.

Wenlandt (1977b) further studied the system K_2O–MgO–Al_2O_3–SiO_2–H_2O–CO_2 up to 30 kb to define the stability of phlogopite with possible implications as to the nature of liquid from the melting of periodotites and kimberlites. He found that, at 20 kb, 1160°C and in the presence of a large amount of water, divariant melting of phlogopite occurs according to the reaction

Phlogopite + Leucite + Kalsilite + vapor → Forsterite + Liquid   (1)

The melting is eutectic-like at high CO_2 and low H_2O content. Above 30 Kb, the melting may be described by the reaction

Phlogopite + Enstatite + Magnesite → Forsterite + Liquid + Vapor   (2)

The reactions (1) and (2) clearly indicate that pressure dependent melting controls the composition of the liquid. The increase in pressures produces a change in the liquid composition from slightly to highly silica undersaturated, maybe even with kimberlitic affinities.

CHAPTER 5

The Solubility of Water and Carbon Dioxide in Silicate Melts

Studies of fluid inclusions in minerals (Roedder, 1965), chemistry of rocks (Nockolds, 1954), and direct observations of volcanic eruptions (Heald et al., 1963; Macdonald, 1972) show that water and carbon dioxide are the most abundant volatiles in igneous rocks (Table I). The importance of volatiles in magma evolution and crystallization is widely recognized. To further understand the role of volatiles in igneous process it is imperative to determine the extent of their solubilities in silicate melts. Considerable research effort has only recently been focused on the solubility measurements of CO_2, H_2O, CO_2 + H_2O in silicate melts under varying pressure and temperatures to quantitatively assess their role in magma crystallization and the melting processes in the mantle.

The solubility measurements of volatiles in silicate melts is carried out by the following techniques:

1 Weight loss method of glass compositions (see Holloway and Lewis, 1974).
2 Phase boundary determination method (see Yoder et al., 1957; Eggler, 1973).
3 Beta-track method for CO_2 using C_{14} (see Mysen et al., 1974).

A major portion of data on the solubility determinations of volatiles, especially CO_2, in silicate melts at high pressures, is from the work at the Geophysical Laboratory, Washington, D.C.

1 THE SOLUBILITY OF WATER IN SILICATE MELTS

The studies by Goranson (1931, 1937, 1938) on the silicate–water system were pioneering for experimental determination of solubility of water in a silicate melt up to a pressure of 4 kb/P_{H_2O} and temperatures of 900–1200°C. He found the

The Solubility of Water in Silicate Melts

solubility of water to be a maximum of 11.82 and 11.0 wt. % at temperatures of 900 and 1200°C, respectively.

The theoretical basis for solubility of one component into another can be expressed by the following relationship (see Turner and Verhoogen, 1960, pp. 416–420; also Shaw, 1964; Holloway et al., 1968; Burnham, 1967):

$$\frac{dN}{dP} = \frac{V^g - \overline{V^s}}{(\Delta\mu/\Delta N)_{P,T}}$$

where N = mol. fraction of water in melt

μ = chemical potential

V^g = mol. vol. of water as gas

$\overline{V^s}$ = partial mol. vol. of water in melt

P,T = pressure, temperature

But for ideal solution

$$\frac{\Delta\mu}{\Delta N} = \frac{RT}{N}$$

therefore by substitution

$$\frac{dN}{dP} = \frac{(V^g - \overline{V^s})N}{RT} \quad \text{or} \quad RT\frac{dN}{N} = (V^g - \overline{V^s})dP$$

now by integration

$$2.303 RT \log N = \int_1^P (V^g - \overline{V^s})dP$$

For ideal solution, according to Goranson, $\overline{V^s}$ is about 0.7 times the volume. The above equation reflects the relationship of absolute solubility of water in silicate melts as governed by physical parameters. The basis of Goranson's (1938) measurements can be expressed by the relationship

$$S_{H_2O} = \frac{P}{a + bP}$$

where S_{H_2O} = solubility of water in weight percent

P = pressure

a and b are constant at any one given temperature.

Therefore, at low pressure S_{H_2O} is proportional to P. When P is large, then S_{H_2O} approaches $1/b$. In other words, there is an upper limit to the solubility of water. It is not possible to have silicate melts with 50% H_2O (Goranson 1937, 1938).

The solubility relationships in a two-component liquid–gas system forming an ideal solution can also be expressed by Henry's Law, which states:

$$N_i = Kp_i$$

where N_i = mole fraction of gas i in its saturated solution

K = constant that is a function of temperature

p_i = vapor pressure of the gas i

It is apparent that the solubility of a gas in a liquid has a direct relationship to its vapor pressure. As the vapor pressure of the gas increases, so does its solubility. Thus in geological environments the solubility of a volatile component should increase with pressure (depth). For some volatile components with large molecular weight or ionic or molecule size there may be a certain critical or minimum pressure required before its solubility can be achieved.

Water dissolves in silicate melts as OH', H_2O, and H_2 and its function is to break the Si–O–Si bridges. The values of solubilities of water in various silicate melt compositions at different pressures and temperatures from many published works (Goranson, 1937, 1938; Hamilton et al., 1964; Yoder et al. 1957; Stewart, 1957; Tuttle and Bowen, 1958; Luth, 1969; Brey and Green, 1977; Yoder, 1954, 1958; Yoder and Kushiro, 1969; Hodges, 1973, 1974; Burnham and Jahns, 1962; Khitarov and Kadik, 1973; and others) are listed in Table XVI from which the following observations can be made:

1 The solubility of H_2O in silicate melts increases with increasing pressure to a maximum value, which is different for different compositions.

Table XVI The Solubility of Water in Selected Silicate Melt Compositions[a]

Melt Composition	Temp. (°C)	Pressure (kb)	Wt. % H_2O
Diopside	1240	20	17.0
	1425		20.2
Diopside	1280	30	21.5
Forsterite	1375	30	27.0
Albite	1000	1	5.5
		2	7.0
		3	8.0
Albite	1200	2	6.6
		3	8.0
		5	11.0
		10	24.0
Nepheline	1200	3	9.0–10.0
Anorthite	1200	3	6.0
Silica	950	15	24.0
Granite	660–820	2	7.0 (6.0)[b]
		4	10.0 (9.0)[b]
		6	12.5 (11.0)[b]
		8	14.0 (13.0)[b]
		10	20.0 (18.0)[b] (22.0)[c]
Syenite–nepheline syenite	850–900	1	~7.50
Basalt	1100	2	4.0
		4	6.0
Basalt	1200	6	7.5
		10	14.0
Andesite	1100	2	6.0
		4	8.0
		5	9.0
Olivine basanite	1200	27	17.0
Melt coexisting with mantle assemblage	1200	20	20.0
$Di_{50}Fo_{30}An_{20}$	1200	3	3.0
		5	4.0
		10	10.0
Olivine melilitite	1110	30	30.0–34.0

[a] Based on data from Goranson (1931–1938), Tuttle and Bowen (1958), Burnham and Jahns (1962), Brey and Green (1977), Hamilton et al. (1964), Luth et al. (1964), Khitarov and Kadik (1973), Hodges (1973, 1974), and Eggler (1975).
[b] At 900°C.
[c] At 1200°C.

2 For a given pressure and temperature, the solubility is higher in salic than mafic melts. At pressures of up to 10 kb/P_{H_2O} and temperatures of 1000 to 1200°C, the granitic melts can dissolve 20.0–22.0 wt.% H_2O whereas only 14.0 wt. % H_2O is soluble in basaltic melts under similar conditions.

3 Under crustal pressure and temperature conditions, the solubility of H_2O in silica-undersaturated (nepheline-bearing) salic melts is higher than silica-oversaturated (quartz-bearing) melts.

4 Under mantle pressure and temperature conditions, olivine basanitic to olivine melilititic melts can dissolve 20.0–30.0 wt. % H_2O (at 27–30 kb and 1100–1200°C).

5 The solubility of H_2O in general seems to depend on the degree of silica saturation of the melt.

6 The solubility of H_2O in ultramafic melts is also very high. Above 20 kb diopside and forsterite can dissolve up to 27.0 wt. % H_2O.

The high degree of solubility of H_2O results in much lower melting temperatures, thus enhancing the possibility of magma generation by partial melting of the mantle material within the pressure and temperature limits defined by the geothermal gradient(s).

2 THE SOLUBILITY OF CARBON DIOXIDE IN SILICATE MELTS

The role of CO_2 in crystallization–differentiation of alkaline carbonatite and ultramafic magmas has been stressed by many workers from field relations and experimental studies of systems containing CO_2 (Wyllie, 1960, 1965; Roedder, 1965; Wyllie and Boetcher, 1969; Wyllie and Haas, 1965; Watkinson and Wyllie, 1964; Wyllie and Tuttle, 1959a; Gold, 1966; Von Eckermann, 1966; Pecora, 1956; Patterson, 1958; Bailey, 1966; Tuttle and Gittins, 1966; Wyllie and Huang, 1976; Smith, 1956; Egorov, 1970; and others). However, there are very few data on the solubility of CO_2 in silicate melts. The low pressure melting relations of granitic rocks has generally indicated that CO_2 does not affect the melting and crystallization curves in the same manner as does water (Wyllie and Tuttle, 1959b). Up to pressures of 10 kb, CO_2 was considered rather insoluble or meagerly soluble in silicate melts (Holloway and Burnham, 1972; Yoder, 1973c; Pearce, 1964; Kostervangroos and Wyllie, 1966; Eggler and Burnham, 1973).

However, Hill and Boetcher (1970) had suggested increased solubility of CO_2 in basaltic melts at pressures above 15 kb.

The work of Eggler (1973)—pioneering in many respects—showed that at high pressures (above 15 kb) an appreciable amount of CO_2 is soluble in silicate melts. This aroused much interest in the determination of solubilities of CO_2 in important silicate minerals and rock melts as a function of pressure and temperature (see Eggler, 1974, 1975a,b, 1976; Mysen, 1975, 1976; Green, 1973; Rosenhauer and Eggler, 1975; Kadik and Eggler, 1975; Boetcher et al., 1975; Wyllie and Huang, 1976; Holloway et al., 1976; Eggler et al. 1974, 1976; Brey and Green, 1976, 1977).

Table XVII lists the available data on CO_2 solubility in silicate mineral and rock melts.

Some pertinent observations are summarized below:

1. The solubility of CO_2 in silicate melts is dependent on pressure, temperature, and the bulk composition of the melt. With increasing pressure the solubility is higher in mafic than salic melts. It is a function of basicity of the melt.

2. At pressures below 15 kb, the solubility of CO_2 in silicate melts is very low, for example, less than 2.0 wt. % in felsic melts and is not related to the type of cation.

3. At pressures of 15–30 kb the solubility of CO_2 in mafic melts sharply increases. Diopside melt can dissolve 4.0–5.0 wt. % CO_2, whereas in olivine-bearing nephelinite and melilitite melts almost 9.0 wt. % CO_2 is soluble at 30 kb.

4. The solubility of CO_2 in silicate melts increases with increasing H_2O content of the melt.

5. The overall CO_2 solubility in silicate melts is much lower than that for H_2O for given pressure and temperature conditions. At 30 kb, 17.0 wt. % H_2O goes into solution in diopside melts compared to only 4.0–5.0 wt. % CO_2.

6. The solubility of CO_2 in melts in equilibrium with mantle mineral assemblage may reach as high as 40.0–45.0 wt. %.

7. CO_2 dissolves in silicate melts mostly as a carbonate (CO_3^{2-}) ion.

The high degree of CO_2 and H_2O solubilities in melts under upper mantle conditions is important in terms of pressure and temperature conditions of magma generation, partial melting in mantle, and possible controls on ultramafic,

Table XVII Solubility of Carbon Dioxide in Selected Silicate Melts[a]

Melt Composition	Temp. (°C)	Pressure (kb)	Wt. % CO_2
Diopside	1625	5	1.2
		10	3.0 (1.68 at 1580°C)
		20	3.3 (2.4 at 1580°C)
		30	4.8
Nepheline	1625	30	4.9
Jadeite			3.3
Albite			2.2
Nepheline	1525	10	1.6 (1.4 at 1450°C)
		20	2.3
		25	2.6
		30	2.8
Jadeite	1525	10	1.5 (1.1 at 1450°C)
		20	1.95
		30	2.30
Albite	1525	10	1.4 (1.12 at 1450°C)
		20	1.7 (1.4 at 1450°C)
		30	1.9 (1.6 at 1450°C)
Plagioclase	1450	30	<2.0
Orthoclase	1625	5	1.48
$Di_{35}Py_{65}$	1500	15	4.1
	1530	20	4.9
$Di_{60}Fo_{30}Sil_{10}$	1575	15	3.0
$Ca_{0.5}AlSi_3O_8$	1625	20	1.49
$Ca_{0.25}Mg_{0.25}AlSi_3O_8$	1625	20	1.61
Olivine melilitite	1450		9.0
	1550	30	8.5
	1650		8.3
	1200	30	12.0–21.0 (with 8.8 H_2O)
Olivine melilitite	1150–1300	30	14.0 (with 10.0 H_2O)
Olivine melilitite (partial melt) with Ol or Ga + Px	1160–1180	27	6.0–7.0 (with 7.0–8.0 H_2O)
Olivine melilitite nephelinite	1625	10	3.5
		20	5.1
		30	7.0
		40	7.7

The Solubility of Carbon Dioxide in Silicate Melts

Table XVII (continued)

Melt Composition	Temp. (°C)	Pressure (kb)	Wt. % CO_2
Melt along Px–carbonate join in CaO–MgO–SiO_2–CO_2	1200	24–25	4.0–5.0
coexisting with mantle minerals	≈1200	>25	Up to 40.0–45.0
Granite	1200	1	0.25–0.35
Basalt	1200	3	0.50–0.60

^aBased on data from Mysen (1976), Mysen et al. (1976), Eggler (1973, 1974, 1975), Brey (1976), Brey and Green (1976, 1977), Wyllie and Huang (1976), Wyllie and Tuttle (1959), Wyllie (1977), and Kadik and Khitarov (1973).

carbonatitic and alkaline magmas (Wyllie, 1977a,b; Wyllie and Huang, 1976; Eggler, 1974, 1975a,b, 1978).

On the basis of carbonation and hydration reaction studies of various compositions, Wyllie (1977), Wyllie and Huang (1975, 1976), Lambert and Wyllie (1968, 1970), Moderski and Boetcher (1973), and Yoder and Kushiro (1969) propose that amphibole, phlogopite, and Ca-Mg carbonates might be the likely phases to store H_2O and CO_2 in the mantle (75–150 km) and also as "aqueous pore fluid." The beginning of melting may cause such volatiles to be further dissolved in the melt.

CHAPTER 6

Melting Relations of Rocks: The Whole Rock System

In the preceding chapters it has been illustrated that the phase equilibrium data in synthetic silicate systems complements many of the conventional petrological interpretations and additionally provides a better physicochemical framework for developing petrogenetic models of magmatic evolution. Phase relations in synthetic silicate systems help elucidate and explain, quantitatively, many intrinsic and extraordinary aspects of magmatic crystallization.

The other approach in experimental petrology incorporates the study of melting relations of igneous rocks under controlled pressure and temperature conditions, using the whole rocks as the starting materials. This approach is based on the assumption that crystallized or partly crystallized rock may be treated as a single bulk composition in a multicomponent system, and the course of crystallization can be followed in the usual way by the quenching method (Shepard et al., 1909; Schairer in Bockris et al., 1959).

It must be pointed out that the stability curves of minerals in melting of whole rocks are generally at lower temperatures than those of the individual minerals. That is why you cannot put together the melting (or stability) curves of individual minerals to represent the melting of rocks in general. This shows the overall importance of other components in the rock system.

There are certain limitations to this method:

1 Differences in kinetics of reactions between the laboratory and natural conditions.
2 The rocks represent an end product of a process, which in nature may have been accomplished in several different ways.
3 Difficulties inherent in the experimental conditions, for example, oxidation state, metastability and attainment of equilibrium state, and so on.

Notwithstanding these limitations, melting relations give estimates of temperatures of beginning of melting, complete melting, melting intervals, and information on crystallization sequences of mineral phases. Such data collectively are useful in assessing stability limits of minerals, locales of magma formation, and physical parameters of partial melting. Subsequently the melting relations of basalts, syenites, feldspathoidal syenites, and phonolites will be described.

1 MELTING RELATIONS OF BASALTS WITH WATER

The studies of Yoder and Tilley (1962) on the melting relations of basalts were pioneering in many respects. They studied the melting relations of selected basaltic rock types between 1 atm and 10 kb/P_{H_2O} to ascertain the temperatures of appearance and disappearance of phases as a function of water vapor pressure. Some general observations of Yoder and Tilley on the melting of basalts in the presence of water are briefly summarized below, and melting data for olivine tholeiite and alkali basalt are shown in Fig. 48a and 48b, respectively.

1. With increasing water vapor pressure the liquidus temperatures are depressed by over 150°C, whereas the beginning of melting decreases by over 300°C. The decrease, however, is greatest below 5 kb/P_{H_2O}.
2. The stability limits of anhydrous phases are also decreased with increasing water vapor pressure. Basalt is unstable above water vapor pressure of 1.4 kb/P_{H_2O}.
3. Based on initial composition, the sequence of the appearance of phases is changed or reversed at high pressures. Amphibole may be the liquidus phase in basaltic compositions at pressures above 12 kb/P_{H_2O}.
4. Plagioclase is generally the last phase to crystallize in all compositions.
5. The principal phases crystallize over a short temperature interval (150°C at 1 atm), and this interval somewhat increases with water vapor pressure. This implies that basalts lie close to four-phase curves.
6. The liquidus temperatures are also within a narrow range (~1000–1100°C at 10 kb/P_{H_2O} for basalt samples studied by Yoder and Tilley, 1962). Measured temperatures of some lava eruptions are listed in Table XX.
7. The melting intervals increase sharply with increasing water vapor pressure.
8. Oxidation state of iron may influence the sequence of iron-bearing minerals.

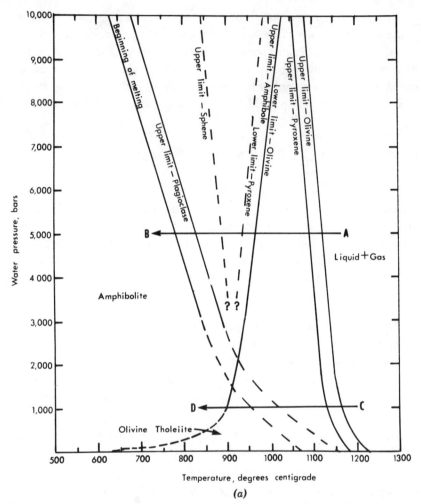

Figure 48 The basalt–water system. (a) The melting relations of olivine tholeiite to 10 kb/P_{H_2O}. Note at 5 kb/P_{H_2O}, the first phase to completely melt is plagioclase, whereas it is an amphibole at 1 kb/P_{H_2O}. (After Yoder and Tilley, 1962. With permission of the *Journal of Petrology*.)

Consider cooling and crystallization of a melt at 5 kb/P_{H_2O} (point A in Fig. 48a). At 1125°C olivine separates and is joined by pyroxene at 1090°C. Between 900 and 960°C both olivine and pyroxene react with gas, and amphibole becomes the stable phase at 960°C. Sphene appears at about 900°C and plagioclase at 850°C. The melt crystallizes at about 800°C giving Am + Pl + Sph + Mt.

Melting Relations of Basalts with Water

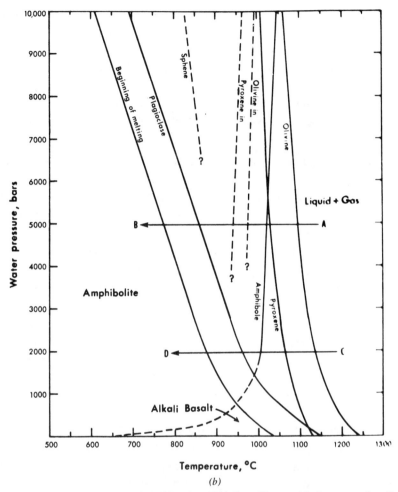

Figure 48 (b) Melting relations of alkali basalt to 10 kb/P_{H_2O}. Note amphibole appears after olivine instead of pyroxene at 5.3 kb/P_{H_2O} and above. Amphibole may be the liquidus phase at ~11 kb. At pressure of about 3 kb/P_{H_2O}, the sequence for plagioclase and pyroxene seems to be reversed. (After Yoder and Tilley, 1962. With permission of the *Journal of Petrology*.)

However, the cooling along line CD (Fig. 48b) at a pressure below 1.5 kb P_{H_2O} will produce a different assemblage. At 1175°C olivine crystallizes followed by pyroxene and plagioclase at 1140°C and 950°C, respectively. The melt will completely crystallize at about 900°C to an Ol + Pl + Px + Mt assemblage—a common basaltic assemblage. As amphibole is stable just above CD, olivine and pyroxene cannot react with the gas phase to form amphibole. Basalt crystallizes at surface or near surface environments so the gas phase is lost. This should explain why amphiboles are not common to basaltic mineralogy. Compare the cooling along lines *AB* and *CD* for alkali basalt in Fig. 48*b*. What differences do you think exist?

Some of the important conclusions from Yoder and Tilley's (1962) investigations of melting relations of synthetic and natural basalts are summarized below:

1. The increased stability of amphibole at high water vapor pressure is suggestive of chemical similarities between basalts and hornblendites. In other words, hornblendites are chemical equivalents of basalts. Therefore, complete melting of hornblendites (or amphibolites) could generate hydrous basalt magmas. This lends credence to Bowen's (1928) suggestion that resorption of hornblende may cause a tholeiite to change to an alkali basalt trend.

2. The partial melting of an amphibolite may produce an anorthositic liquid at low temperatures (600°C), as plagioclase is the first phase to melt at pressures above 1.5 kb/P_{H_2O}.

3. The steps of the reaction series, particularly in the discontinuous part, may not follow the scheme set by Bowen (1922) at pressure above 10 kb/P_{H_2O} or below 2 kb. However, between 2 kb and 10 kb, olivine and pyroxene react with liquid and gas to form amphiboles. Yoder and Tilley (1962, p. 467) suggest that "only the tholeiites would be capable of exhibiting the steps in the series."

4. Every basalt at high pressure has an equivalent eclogite. In contrast to basalts, eclogites have rather a short melting interval (<100°C), and both pyroxene and garnet are liquidus phases. Thus eclogites may be a depth-dependent fractionally crystallized product of liquid obtained from partial melting of more primitive material—garnet peridotite. Therefore at high pressures removal of garnet will enrich the liquid in omphacite—an alkali basalt trend, whereas the removal of omphacite will enrich the garnet component in the liquid—a tholeiite trend.

    In other words, pressure–dependent fractionation or melting could form different basalts. Therefore depth and degree of partial melting exert important controls on basalt compositions.

5. Alkali basalt melt will form at a greater depth than a tholeiite melt from the same primary source. The course of melt fractionation of the principal basalt series are thus defined in the region of magma generation.

6 Equilibrium crystallization of one basalt magma can not produce both silica-undersaturated (Ne-bearing) and silica-oversaturated (Qtz-bearing) derivatives at the same time, an indication of a thermal divide at low pressure.

7 Based on the short temperature interval for the appearance of major phases, it is suggested that magma compositions lie close to the univariant lines or invariant points.

8 The shifts in oxygen pressure may influence the liquid composition to trend from silica-undersaturated to –oversaturated.

9 Eclogite could change to a gabbro or pyroxenite with increasing temperature and to an amphibolite or pyroxene hornblendite, if water is available at depth.

The melting work on basalts and ultramafic rocks has been extended by Tilley et al., (1964, 1967, 1968), Tilley and Yoder (1964), Thompson and Tilley (1969) and more recently by Mysen (1973, 1975), Green (1975), Mysen and Boetcher (1975), and Arndt (1976, 1977). The large volume of data establishes that liquidus temperatures of mafic volcanic rocks are inversely related to the iron-enrichment index $FeO + Fe_2O_3/FeO + Fe_2O_3 + MgO$. The cumulate rocks have higher liquidus temperatures in general and low iron-enrichment indices. The high liquidus temperatures may also be related to the primitiveness of the material. This may then be linked to the source of the parent and the depth of formation.

Osborn (1963) and Hamilton et al., (1964) from the melting relations of basalts under controlled oxygen and water vapor pressure conditions suggested that in addition to the change of phases at the liquidus and subliquidus temperatures, there is also a wide temperature range over which magnetite appears. The temperature for the appearance of plagioclase and pyroxenes decreases with decreasing oxygen pressure. They concluded that high P_{O_2} favors a silica-enrichment trend—a possible mechanism to derive andesites by fractional crystallization of basalts at depth.

2 MELTING RELATIONS OF BASALTS WITHOUT WATER

Green and Ringwood (1964, 1967a) published a detailed account of studies on the melting behavior of basalts up to anhydrous pressures of 30 kb. Their data on olivine tholeiite and alkali olivine basalt are presented in Fig. 49a and b, respectively. Their studies bring to light many points related to behavior of basalts with pressure. Some features related to the principal phases are summarized below:

1. Olivine is a primary phase only at pressures below 10 kb, implying a contraction in the stability field of olivine with pressure.
2. Pyroxenes (aluminous Opx and Cpx) are principal stable phases in the pressure range of about 12–18 kb. In the pressure range of the experiments clinopyroxenes are more stable at high pressures.
3. Plagioclase feldspar disappears as an essential phase at pressures above 10 kb and is replaced by spinel or garnet.
4. Garnet is the liquidus phase at pressures above 20 kb.
5. The melting interval of about 275°C at 1 atm is narrowed to about 50°C at pressures of 30 kb for an olivine tholeiite. However, the melting interval of alkali olivine basalt is not that drastically affected with increase in pressure.
6. The short temperature interval (50°C) over which Ga + Cpx crystallize is in conflict with Yoder and Tilley's suggestion that prolonged high pressure fractionation of garnet or omphacite could yield alkali-basaltic or tholeiitic derivatives.

On the basis of their data Green and Ringwood (1967) proposed a pressure-dependent fractionation scheme for basalts, which is shown in Fig. 50.

A parent magma of olivine-rich tholeiitic composition has the potential of producing different derivative magmas, through fractionation, as a function of pressure.

At deeper levels (45–65 km depth) aluminous Opx + Cpx are the dominant phases as the field of olivine crystallization highly contracts. The fractionation of the orthopyroxene (with or without clinopyroxene) phases may push the liquids toward critical silica-undersaturation characterizing an *olivine tholeiite→olivine basalt→alkali olivine basalt→olivine basanite* trend (see Fig. 50; cf. Yoder and Tilley, 1962).

At depth of about 30 km the field of olivine crystallization is somewhat less contracted. Therefore the fractionation of Ol + low Al–En ± Cpx will result in high Al_2O_3 basaltic derivatives such as (also see Fig. 50):

1. Olivine-rich tholeiite→high Al_2O_3 olivine tholeiite.
2. Low Al_2O_3 olivine basalt→high Al_2O_3 alkali olivine basalt.
3. Low Al_2O_3 alkali olivine basalt→high Al_2O_3 alkali olivine basalt (Ne-normative).

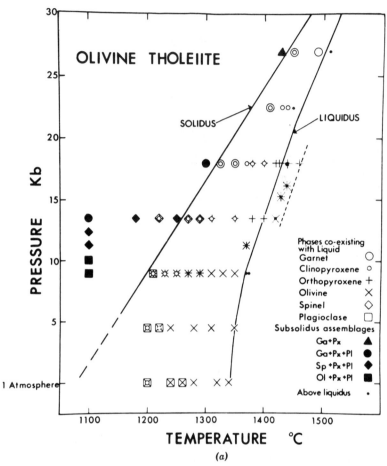

Figure 49 Anhydrous melting of basalts at high pressures. (a) Anhydrous melting of olivine tholeiite to pressures of 30 kb. Note the melting interval narrows with increasing pressure. The dominant liquidus phase is olivine below 10 kb, orthopyroxene between 10 and 20 kb. (After Green and Ringwood, 1967a, and Green, 1970. With permission of the *Contributions to Mineralogy and Petrology*.)

At depth of 15 km or less, the field of olivine crystallization is dominant and the fractionation of olivine may push the liquid toward silica-oversaturation giving olivine tholeiite → quartz tholeiite trend.

It may be concluded from the above summary that the thermal divide in basalts at low pressures may be inoperative at high pressure.

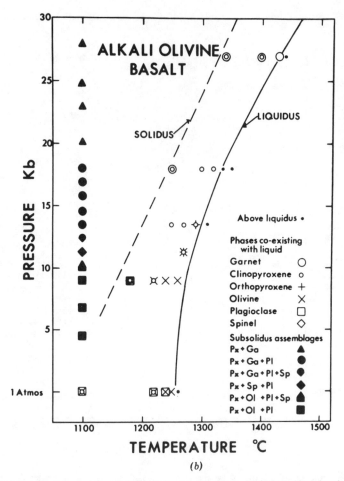

Figure 49 (b) Anhydrous melting of alkali olivine basalt to pressure of 30 kb. Note that the melting interval is about 100°C for most of the pressure range. The dominant liquidus phase is olivine below 10 kb, clinopyroxene between 10 and 20 kb, and garnet above 25 kb. Therefore, at 10–20 kb, pressures orthopyroxene (and clinopyroxene) fractionation would be important determinants of liquid trends. (After Green and Ringwood, 1967a. With permission of the *Contributions to Mineralogy and Petrology*.)

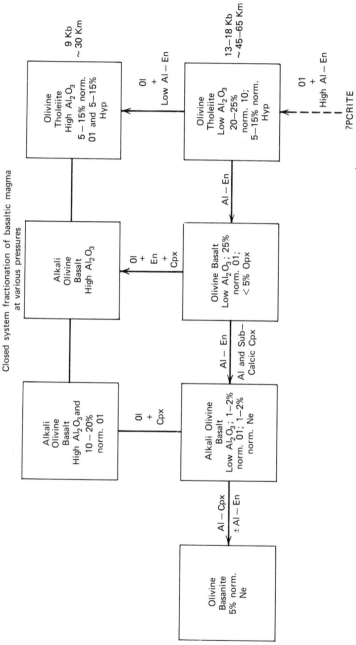

Figure 50 Fractionation scheme for basaltic magmas at various pressures. (After Green and Ringwood, 1967a. With permission of the *Contributions to Mineralogy and Petrology*.)

The melting work on basalts has facilitated quantification of basaltic fractionation. Importantly, it has suggested that major basalt types could be produced through fractionation of appropriate phases in a very narrow pressure or depth interval of 30–70 km.

3 MELTING RELATIONS OF PHONOLITES AND AGPAITIC* SYENITES AND NEPHELINE SYENITES

Melting relations of felsic alkaline rocks of diverse compositions from alkaline complexes of Kenya, Africa, Ilimaussaq and Grondal-Ika, South Greenland, Tanrief Volcanics, South Arabia, and Reunion Islands have been studied under anhydrous 1 atm pressure conditions (Piotrowski and Edgar, 1970; Sood and Edgar, 1970; Humpheries and Cox, 1972; Humpheries, 1972), water vapor pressure of up to 2 kb (Barker, 1965; Sood and Edgar, 1967; Sood and Edgar, 1970) and under controlled partial pressure of oxygen using nickel–nickel oxide and magnetite–hematite buffers at water vapor pressure of 1 kb (Sood and Edgar, 1970). Figure 51a–d shows the melting curves for a selected phonolite, syenite, nepheline syenite, and agpaitic nepheline syenite. Melting studies of felsic alkaline† rocks show the following general relationships:

1   The increase in water vapor pressure depresses the liquidus temperatures by as much as 300°C compared to liquidus at 1 atm. The enhanced depres-

---

*Ussing (1912) introduced the term agpaite to describe a subgroup of nepheline syenites, which make up the Layered Series of the Ilimaussaq batholith, South Greenland. According to Ussing these agpaitic rocks are distinguished from other nepheline syenites by the following characteristics:

1   There is an excess of alkalis in proportion to alumina according to the formula $Na_2O + K_2O/Al_2O_3 \geq 1.2$.
2   The Fe^{3+} content is high and is mainly present in aegirine, arfvedsonite, or aenigmatite and not as iron ore.
3   Mg and Ca are low.
4   A characteristic feature is the presence of Zr and Cl, which may occur in considerable quantities (0.06–3.10 wt. % Cl), mainly in eudialite and sodalite.

Ussing (1912) originally restricted the term "agpaitic" to nepheline syenites. Goldschmidt (1930) and Polanski (1949) extended the term to acid peralkaline rocks containing zircon and/or pyrochlore, which are absent from the agpaitic rocks. However, Gerassimovsky (1956), on the basis of available data, proposed that the ratio $Na + K/Al \geq 1$ (in mol. prop.) be used to define the agpaitic character of rocks (also see Sorensen, 1958, 1960, 1974).

†See Merril et al. (1970) for effects of up to 20 kb/P_{H_2O} on the melting behaviors of felsic rocks.

Melting Relations of Phonolites—Syenites

sion of liquidus may be due to increased solubility of water in the melt. The first kilobar, however, has the greatest effect in all rocks (see Fig. 51a–d).

2. The temperature at which primary phase appears for different compositions lies within narrow ranges, 1125–1198°C dry, and 870–910°C under water vapor pressure conditions.

3. The primary phase at the liquidus for most of the rocks is either feldspar or clinopyroxene with the exception of volatile-rich naujaite in which

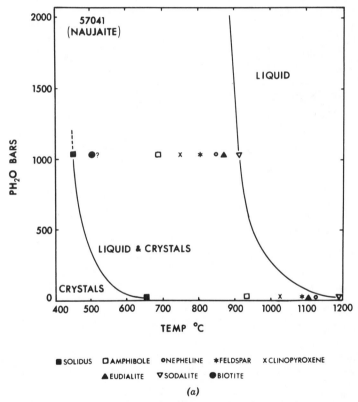

Figure 51 Melting relations of felsic rocks to 2 kb/P_{H_2O}. Note that with increasing P_{H_2O}, liquidus temperatures are depressed substantially and appearance of phases changed. Also, the melting intervals are related to the mode of formation and the contents of the volatile bearing minerals. (After Sood and Edgar, 1970. With permission of the *Meddlesser om Groenland*.) (a) Naujaite–water system.

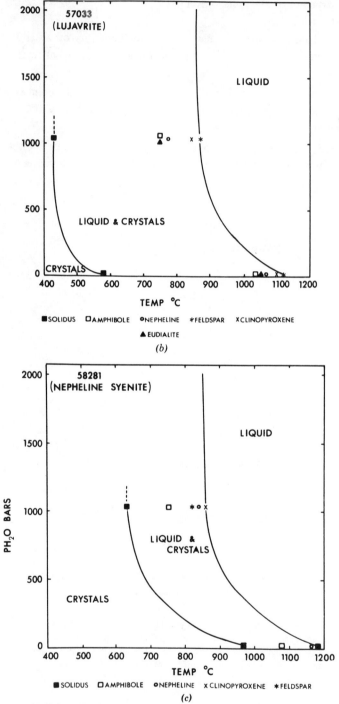

Figure 51 (b) Lujavarite–water system. (c) Nepheline syenite–water system.

Melting Relations of Phonolites—Syenites

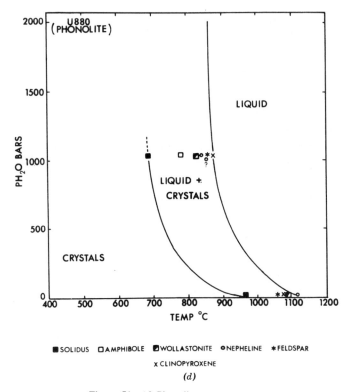

Figure 51 (*d*) Phonolite–water system.

sodalite is the primary phase. Trachytes have plagioclase as the liquidus phase (Humpheries and Cox, 1972).

4 There is no change in the liquidus phase at 1 or 2 kb/P_{H_2O}.

5 The change of phase from feldspar or nepheline at 1 atm to clinopyroxene in the presence of water was observed for some rocks. This may be related to increased depression of liquidii for felsic relative to mafic minerals (see Table XIX).

6 The crystallization sequence of the principal minerals in most rocks is either feldspar–nepheline–clinopyroxene or clinopyroxene–feldspar–nepheline. These phases tend to crystallize over a short temperature interval (30–50°C) for rocks with small volatile content. This suggests that compositions of such rocks must lie close to the four-phase equilibrium or fractionation curves.

7   Amphibole is usually the last mineral to appear in all the rocks both under 1 atm and elevated water vapor pressure.
8   The melting intervals appear to be directly related to the contents of volatile-bearing minerals in the rocks and agpaitic character (see Table XVIII). Naujaite, a sodalite-bearing syenite, has the largest interval (>400°C).
9   Volcanic rocks melt over a smaller temperature range than the intrusive rocks.
10 The temperature of the beginning of melting seems to be related to the volatile contents and agpaitic indices. The higher the volatile content and agpaitic index, the lower is the beginning of melting temperature, for example, phonolite (Fig. 51d), and naujaite (Fig. 51a).

Table XVIII   Comparison of Modal Content of Volatile-Bearing Minerals, Agpaitic Indices and Melting Intervals at 1 kb/P_{H_2O}[a]

Rock	Modal Volatile-Bearing Minerals	Agpaitic Index Na + K/Al	Melting Intervals (°C)
Naujaite (57041)	49.0	1.31	460
Foyaite (57070)	10.2	1.07	365
Lujavarite (57033)	19.0	1.47	450
Foyaite (27113)	8.10	1.05	260
Nepheline Syenite (58281)	7.10	0.95	230
Foyaite (K16)	4.0	0.99	215
Nepheline Syenite (K24)	4.0	1.08	240
Phonolite (U880)	5.0	0.80	185

[a] From Sood and Edgar (1970).

Table XIX   Liquidus Temperatures in °C and Primary Phases Under Different Experimental Conditions[a]

Rock	Liq 1 Atm/ Primary Phase	Liq Argon/ Primary Phase	Liq 1030 Bars P_{H_2O}/ Primary Phase	T = Liq 1 Atm Liq 1030 Bars P_{H_2O}	Norm Alk-Fels + Ne + Lc + Ns	Liq 1030 Bars P_{H_2O} with NNO Buffer	Liq 1030 Bars P_{H_2O} with HM Buffer
Naujaite	1195/Sod	1195/Sod	910/Sod	285	72.50	—	—
Foyaite	1125/Fels	—	885/Fels	240	68.57	855/Cpx	860/Cpx
Lujavarite	1125/Fels	—	880/Fels	245	66.06	860/Cpx	870/Cpx
Foyaite	1185/Cpx	—	875/Cpx	310	75.03	—	—
Nepheline syenite	1155/Ne	1125/Ne	885/Cpx	270	70.60	—	—
Foyaite	1175/Cpx	1165/Cpx	885/Cpx	290	75.11	—	—
Nepheline syenite	1120/Cpx	1095/Cpx	875/Cpx	235	62.71	—	—
Phonolite	1123/Ne	1110/Cpx	885/Cpx	238	67.49	—	—

[a]After Sood and Edgar (1970). Compare liquidus at 1 Atm with lava temperature, Table XX.

3.1   Chemical Parameters vs Liquidus Temperatures

Chemical variation diagrams are used to establish the genetic relationships of the rocks. Chayes (1960, 1962) pointed out that certain limitations exist in the classical variation diagrams where the percentage data, which form closed arrays that is, "a situation where variables plotted are often percentages of the same whole" (Chayes, 1962, p. 440) and that "there is strong bias toward negative correlation between variables of closed arrays" (Chayes, 1962, p. 442). He suggested that "the effects of such a closed array may be reduced or eliminated by relating the restricted variables to an outside variable" (Chayes, 1960, p. 4185). A good correlation exists between liquidus temperatures and SiO_2, Al_2O_3, TiO_2, contents and $FeO + Fe_2O_3/FeO + Fe_2O_3 + MgO$—the iron-enrichment indices of alkaline rocks from Greenland and Kenya, Africa (Fig. 52). The iron-enrichment index defines progressive differentiation, particularly in the Kenyan rocks (see Tilley et al. 1964, for similar relations in basalts).

3.2   Relationship Between Melting Intervals, Volatile Contents, and Agpaitic Indices

From the melting data on the phonolites, agpaitic syenites, and nepheline syenites from Kenya, Ilimaussaq, and Grondal-Ika, South Greenland, Sood and Edgar (1970) reported that there exists a direct relationship between the melting intervals, volatile contents, and the agpaitic indices of the rocks studied. Rocks with

Table XX Observed Temperatures of Lavas

Locality	Year	Temp. (°C)	Lava Type	References
Kilauea, Hawaii	1909	1000	Basalt surface	Daly, 1911
	1952	1095–1130	Basalt lava	MacDonald, 1953
	1963	1140	Basalt lava	Peck et al., 1964
Nyiragongo	1948–1959	987–1080	Nephelinite	Sahama and Myer, 1958, and Tazief in MacDonald, 1972
Nyamuragira		1095	Leucite basalt	Verhoogen, 1948
Vesuvius Lava	1904	1100	Leucite tephrite	Brun (from Gutenberg, 1951)
	1929	1150		Rittman, 1962
	1913	1200	Leucite tephrite	Perret, 1924
Nyamlagira	1938	1040–1075		Verhoogen, 1939
Santa Maria	1940	725	Dacite	Zies, 1941
Paricutin	1945	1135		Bullard, 1947
Klyuchevskaya	1938	865		Popkov, 1946
Oshima	1950–1951	1060–1125		Minakami and Sakuma, 1951

high modal percent of volatile-bearing minerals (mainly Cl, F) and high agpaitic indices tend to have longer melting intervals, in contrast to those with low modal percent of volatile-bearing minerals and low agpaitic indices (Table XVIII, Fig. 53).

The probable explanation for such long melting intervals lies in the fact that the average content of volatiles, Cl, F, P, and S, alkalis, and rare earths is higher in agpaitic rocks (see Gerassimovsky, 1956, 1965) than in any other type of magmatic rocks. It is believed that the volatiles Cl, F, P, and S are mainly present in the liquid rather than in the gaseous phase in an agpaitic melt, especially those melts having an excess of Na with respect to Al. Thus the volatiles take part in the progressive crystallization of the melt and are "fixed" in the sodium-rich minerals sodalite, eudialite, and arfvedsonite. The main effect of volatiles such as Cl, F, and H_2O, if they do not participate as the structural or compositional units of crystallizing phases, is to lower the liquidus and crystallization temperatures. However, if these volatiles are "fixed" in the principal solid phases (sodalite, eudialite, and arfvedsonite) during the magmatic stage of crystallization, then, in addition to lowering the crystallization and melting temperatures, they extend the range of crystallization of the melt.

A further support to the hypothesis that volatiles such as F, Cl, S, and P are present in the liquid phase, rather than in a gaseous phase, is accorded by the observations of Kogarko and Rhyabchikov (1961). They showed, from calculations of the constants of exchange reactions involving silicates, fluorides and chlorides and on the basis of silicate/halide systems, that separation of volatile

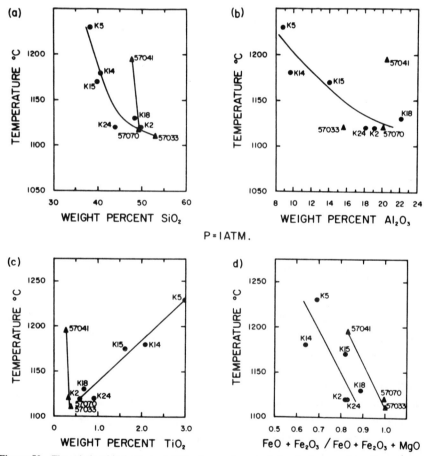

Figure 52 The relationship between liquidus temperatures and chemical composition for various feldspathoidal rocks from Kenya, Africa (denoted by K) and Greenland. (After Sood and Edgar, 1970, and Piotrowski and Edgar, 1970. With permission of the *Meddlesser om Groenland*.)

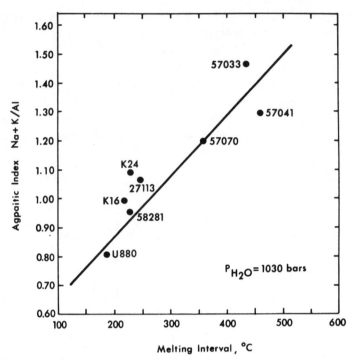

Figure 53 The relationship between agpaitic index and melting intervals for felsic rocks from Kenya, Africa (shown by prefix K) and Greenland at 1 kb/P_{H_2O}. (After Sood and Edgar, 1970. With permission of the *Meddlesser om Groenland*.)

components into the gaseous phase becomes more difficult as the alkalinity of the melt increases because the activity coefficients of Cl, and F are much lower in alkaline melts than in acid ones. This also explains why griesens are commonly present around granites but are less common around nepheline syenites. The relatively short melting intervals observed in miaskitic rocks may be related to the volatile contents. The miaskitic rocks have only H_2O as the important volatile compared to the agpaitic rocks, which contain appreciable amounts of Cl, F, S, and P in addition to H_2O and CO_2 and an initial excess of alkalis, principally Na to Al, in such melts. Sood and Edgar (1970) observed the order sodalite–eudialite–amphibole similar to the natural sequence and contents of respective volatiles. Minor sodalite has been reported in later members (Sorensen, 1958; Ferguson, 1964).

In the silica-undersaturated agpaitic rocks, where (Na + K/Al) ≥ 1, K is mainly fixed in feldspars. The excess of Na is partly bound in Ti- and Zr-silicates and to a lesser extent in Na-amphiboles and aegirine. More often Na combines with Cl to form "activated" NaCl ion and hence unites with nepheline to form sodalite in the early stages of crystallization of the magma, according to the reaction:

$$3NaAlSiO_4 + NaCl \rightleftharpoons Na_3Al_3Si_3O_{12} \cdot NaCl$$

This explains the crystallization of sodalite in the early member of Ilimaussaq complex. Such a reaction relation will result in the depletion of Cl in the melt and diminish the quantity of sodalite in the subsequent members. Instead, the later members are enriched in F, due to difficulty of its fixation in the early minerals nepheline and feldspars, which are more common in the earlier members of the series. Moreover, it is the general tendency of F to be enriched in the rest-magmas (Kograko and Gulyaleva, 1965). Melting studies show good correlation in the sequences and temperatures of appearance of volatile-bearing minerals with volatile contents of agpaitic rocks.

These features indicate that probably Cl and F are more important than CO_2 in the genesis of undersaturated alkaline rocks in general and agpaitic rocks in particular. Gerrassimovsky (1965), from his studies of the Lovozero alkaline massif, concluded that CO_2 is perhaps unimportant in the P–T condition of the formation of agpaitic rocks.

3.3 Combined Effect of Controlled Oxygen and Water Vapor Pressure on the Melting of Undersaturated Alkaline Rocks

The role of varying pressures of oxygen on the crystallization in synthetic silicate systems and basaltic rocks has been presented previously. Fudali (1965), Hamilton et al. (1964), and Hamilton and Anderson (1967) noted that combined water vapor pressure and P_{O_2} were effective in lowering the fusion temperatures and changing the order of crystallization of Columbia River basalts. Sood and Edgar (1970) in studying the crystallization of two iron-rich agpaitic rocks under controlled water vapor pressure and P_{O_2} conditions found that such conditions not only depress the liquidus and the temperature of the appearance of feldspars but also change the sequence of crystallization. Such an effect has been demonstrated in simpler systems (Yoder, 1954, 1958) but the demonstration of this effect in natural rocks is important to establish close similarities in the cooling behavior of synthetic and natural melts (see Hamilton and Anderson, 1967).

One petrological implication of early crystallization of clinopyroxene under controlled P_{O_2} conditions is that it will enrich the residual liquid in alumina in comparison to the early crystallization of feldspars or nepheline. Thus in an originally undersaturated magma, if differentiation can take place in the short temperature interval between the crystallization of clinopyroxene and that of feldspars or nepheline, the crystallization would follow a miaskitic trend. In contrast, the early crystallization of nepheline or feldspars would deplete the liquid in alumina and tend to produce an agpaitic trend. Therefore, a high P_{O_2} may favor an agpaitic trend, and a low P_{O_2} a miaskitic trend.

It is probable, however, that P_{O_2} in the magma chamber is not constant. A possible fluctuation of conditions in the chamber would result in the formation of pyroxene-rich bands (under low P_{O_2}) and feldspar-rich bands (under high P_{O_2}). Ferguson (1964) has reported thin feldspar-rich and pyroxene (aegirine)-rich bands in the lujavarites of Ilimaussaq.

Thus, high P_{O_2} conditions initiate an alumina-deficient, peralkaline liquid trend. According to Bailey and Schairer (1966), a silica-undersaturated melt is not an essential prerequisite to an undersaturated peralkaline trend, if iron enters feldspars. The tendency for iron to enter feldspar probably increases in such a liquid. It is believed that a small amount of substitution of Fe^{3+} in feldspar possibly under a high P_{O_2} condition can also result in a peralkaline silica-undersaturated melt from a slightly saturated or oversaturated parent magma.

3.4 Melting Relations vs Field Relations

A strong similarity and correlation has been observed between experimentally determined and petrographically deduced courses of crystallization in felsic rocks. Sorensen's (1969) observation on rhythmic layering and sequence of crystallization are on a par with the experimental studies.

The appearance of nepheline or pyroxene or feldspar as a primary phase in many rocks is similar to the observation of Dawson (1962) on rapidly cooled volcanic rocks which contain phenocryst of either of these phases. Moreover, such compositions must lie close to the four-phase curves.

This is the observation common to melting data on rocks from different plutonic and volcanic environments with different chemical and field characteristics implying that some fundamental process must play a dominant role in the generation of alkaline rocks and a likely candidate seems to be the crystal-liquid process, which may be assisted by gas transfer and liquid fractionation to some extent.

CHAPTER 7
Magma Generation

At this point, it is appropriate to present a discussion on certain aspects of magma generation as a logical culmination to the description of crystallization and melting relations in synthetic and whole rock systems. The phase relations in silicate systems, in many ways, complement the crystallization behavior of synthetic and natural melts at various low pressures. Experimental studies of silicate systems clearly establish that the low melting fractions in multicomponent silicate systems have chemical characteristics similar to residual liquids derived through crystallization–differentiation of a basaltic magma. Bowen (1928) had long advocated that a magma of basaltic composition is seemingly a parent magma for many igneous rocks, as field, chemical, and petrographic gradations could be readily traced to it in many petrologic complexes. However, there is petrologic and experimental basis for the inference that many granites (Sederholm, 1926; Eskola, 1932; Ramberg, 1944; Read, 1957; Tuttle and Bowen, 1958; Luth et al., 1964; Raguin, 1965; Marmo, 1967; Misch, 1968; Hyndman, 1969) and calc-alkaline rocks (Holmes, 1932; Turner and Verhoogen, 1960; Yoder, 1969; Gilluly, 1971; McBirney, 1969; Green and Ringwood, 1966; O'Hara, 1965; Vitiliano, 1971; Kuno, 1959; Kushiro, 1969a, 1974) may form, independent of differentiation, from a basaltic parent magma.

The existing data on the mineralogy, chemistry, iron-enrichment ratios, and field and melting relations, point to the fact that many other igneous rocks could be produced through crystallization–differentiation of basaltic magmas. In this chapter we consider a parent magma to be of basaltic composition and discuss the models for its generation.*

1 SOURCE MATERIALS

Basalt is the most dominant volcanic rock that has repeatedly reached the surface of the earth in an all "liquid" state in large quantities throughout the geological

---

*For a detailed and excellent treatment, see Yoder (1976).

time. Alkali-olivine basalts (Ne- and/or Ol-normative) tholeiites (Ol- or Hyp-normative) and quartz tholeiites (Hyp–Qtz-normative) are principal compositional basalt types that are commonly recognized. In general terms, they have compositional continuity and a rather restricted chemical (see Table XXI) and mineralogical (see Table II) composition, both in space and time, despite the diversities in the environments of their formation and enormity of the volume of their flows (see Table XXVIII).

Now the question arises, what is the source material that upon melting has the capability of producing basalt magma types?

Daly (1925) proposed subcrustal tachylite (basaltic glass) layer as the possible source. The studies of Yoder and Tilley (1962) showed that basaltic magma may be derived by melting of amphibolite, hornblendite, or eclogite sources (see also Ringwood, 1969). However, Lambert and Wyllie (1970) demonstrated that amphiboles are unstable above 30 kb pressure. Eclogites have higher melting temperatures at depth (100 km) than the host peridotitic material (O'Hara and Yoder, 1967). It should also be noted that the density and chemical composition of the sources mentioned above is similar to that of the melt to be produced. Therefore, almost *complete melting* of the source becomes a prerequisite. However, there

Table XXI Average Chemical Compositions of Different Kinds of Flood Basalts (in wt. %)[a]

	Continental Tholeiites		Oceanic Tholeiites		Alkali Olivine Basalts	
	Average	Range	Average	Range	Average	Range
SiO_2	50.7	44.35–54.60	49.3	42.8 –52.56	47.1	41.04–51.4
TiO_2	2.0	0.9 – 3.99	1.8	0.35– 3.69	2.7	0.92– 4.52
Al_2O_3	14.4	12.48–16.32	15.2	7.3 –22.3	15.3	10.11–26.26
Fe_2O_3	3.2	0.95– 7.56	2.4	0.69– 7.90	4.3	0.53–15.85
FeO	9.8	4.18–13.60	8.0	2.87–13.58	8.3	0.48–13.63
MnO	0.2	0.10– 0.3	0.17	0.09– 0.44	0.17	0.06– 0.36
MgO	6.2	3.52–11.16	8.3	4.59–26.0	7.0	2.66–17.87
CaO	9.4	7.45–11.8	10.8	6.69–14.1	9.0	6.81–14.46
Na_2O	2.6	1.8 – 3.47	2.6	0.90– 4.45	3.4	1.35– 4.8
K_2O	1.0	0.19– 1.74	0.24	0.04– 0.70	1.2	0.13– 2.5
P_2O_5	—	0.09– 0.81	0.21	0.06– 0.56	0.41	0.09– 0.93

[a]From Hyndman (1972).

Source Materials

are physiochemical difficulties in achieving complete melting. Such observations lead to discounting tachylite, amphibolite, hornblendite, and eclogite as possible principal and/or primary source materials for the generation of basaltic magmas. There is a consensus that basaltic magmas must, therefore, form by partial melting of denser and primitive material of the mantle. Bowen (1928) had proposed such material to be of the chemical characteristic of a peridotite.

The major prerequisite of the source seems to be that it must be able to generate a liquid of basaltic composition or a liquid + crystal mixture capable of producing a basaltic liquid at or near surface conditions (Yoder, 1976).

The occurrence in basalts and kimberlites of high pressure dense ($\sim D$ = 3.3) peridotitic, lherzolitic, pyroxenitic, and eclogitic xenoliths have led to the suggestion that such xenoliths represent mantle-derived materials (see Wagner, 1928; Ross et al., 1954; Forbes and Kuno, 1965; Campbell 1977; Kuno and Aoki, 1970; Harris and Middlemost, 1970; Hutchison et al., 1970: Dachin and Boyd, 1976; Boyd and Nixon, 1976). The xenoliths in kimberlites from Africa and Siberia are dominantly garnet-bearing peridotite/lherzolite (Fo 64.0%, Opx 27.0%; Cpx 3.0% and Py 6.0%) assemblage (see Nixon et al., 1963; Dawson, 1968, Von Eckerman, 1967; McGregor and Carter, 1970; Boyd, 1973, 1974). In contrast, the xenoliths in basalts are dunitic to spinel-bearing lherzolitic/peridotitic, though the latter are more common (Ross et al., 1954; Green and Ringwood, 1963, 1967a,b; McGregor, 1968; Kushiro and Aoki, 1968; Carter, 1970; Green, 1970). Kuno (1964) and O'Hara (1967, 1968) consider the xenoliths as crystal fractionates in basalts formed at depth. However, the Mg/Mg + Fe ratios of greater than 0.80, and generally high pressure and temperature equilibration values ($\approx 1300°C$ at 41 kb) for xenolith minerals (see Akella and Boyd, 1974) are to a degree in accordance with those established for upper mantle conditions.

Of the many geophysical determinations made to establish the compositional nature of the upper mantle, Poisson's ratio,* density calculations from the surface wave data, and Monte Carlo inversion methods provide the most useful information.

Poisson's ratios calculated for upper mantle range between 0.245 and 0.260

---

*Poisson's ratio $= \dfrac{1}{2} \cdot \left[\dfrac{(V_P/V_S)^2 - 2}{(V_P/V_S)^2 - 1} \right]$

where V_P = velocity of P waves and V_S = velocity of S waves.

(Doyle and Evringham, 1964), compared to the value of 0.245–0.255 for forsteritic olivines (Verma, 1960; Graham and Barsch, 1969).

The inversion of surface wave data gives a density of 3.33 gm/cc for the mantle and 2.88 gm/cc for the crust (Dorman and Ewing, 1962).

The Monte Carlo-inversions give an average density of 3.48 gm/cc (Press, 1970) and 3.41 gm/cc (Wang, 1972) for mantle up to a depth of 350 to 400 km.

The above considerations suggest that the xenoliths found in basalts and kimberlites fit their upper mantle derivation. This implies that the upper mantle, to a depth of 400 km, is most likely peridotitic (Yoder, 1974, 1976) or pyrolitic (Ringwood, 1969, 1975) with Mg/Mg + Fe ratio of 0.88 or higher. It is certainly not wholly eclogitic.

It is widely accepted that basaltic magmas are generated by partial melting* of undepleted† upper mantle materials. It may be worthwhile to summarize some of the important characteristics of mantle composition models.

1.1 Garnet Peridotite Model

Yoder (1974, 1976) has proposed that garnet peridotite may be considered as a major mantle material capable of producing various basaltic magmas by partial melting, as a function of depth (also see Hess, 1938; Bowen, 1947).

Garnet peridotites are principally composed of four minerals: olivine, orthopyroxene, clinopyroxene, and garnet (pyrope). According to Rickwood et al. (1968) peridotites, on the average, contain less than 15% garnet in comparison to eclogites, which contain more than 30% garnet. Mineralogical and chemical characteristics of garnet peridotite and related rocks are shown in Fig. 54 and Table XXII, respectively. This model composition has some interesting aspects.

It is stable under mantle conditions. Kushiro and Yoder (1966) from their

---

*In the *partial melting* process of a solid there is an equilibrium between the liquid formed and the remaining solid. Partial melting is similar to batch melting, and liquid can be removed in one batch or as it is formed. In the *fractional melting* process of a solid the liquid is removed from the remaining solid as it is formed. The composition of the liquid at each stage of removal is different than the previous liquid, and no equilibrium between solid and liquid exists. The first liquids from partial (or batch) and fractional melting is compositionally similar, but later liquids are very different. These may be important processes in the mantle. (Ringwood, 1975; Yoder, 1976.)

†The term "undepleted" refers to parent material from which low temperature chemical components capable of yielding liquids of basaltic chemistry upon melting have not been fractionated, removed or extracted (cf. depleted) (Ringwood, 1975; Yoder, 1976).

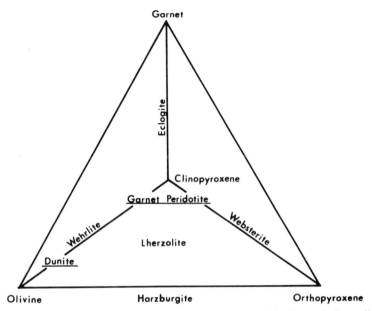

Figure 54 Rock names for ultramafic rocks in the compositional tetrahedron olivine–orthopyroxene–clinopyroxene–garnet. Those underlined plot in the tetrahedron. (Yoder, 1976. With permission of the National Academy of Sciences.)

studies on peridotitic compositions reported that at 1200°C and pressure of 5–30 kb (anhydrous) they show a change from (Fig. 55):

plagioclase peridotite → spinel peridotite → garnet peridotite

as a function of increasing pressure (cf. pyrolite compositions). This may explain the increase in seismic velocities as a function of depth (garnet peridotite is a stable assemblage at 12.5–30 kb). More importantly, such a change would suggest that plagioclase would be at least a normative component of liquid produced from partial melting of undepleted peridotitic compositions. Kushiro (1973) found that the glass obtained from partial melting of garnet lherzolite has normative plagioclase character but not a plagioclase phase.

The studies in the anhydrous systems composed of various combinations of minerals (forsterite, diopside, enstatite, garnet) of peridotites have shown the presence of piercing points at different pressures but compositionally similar to major basaltic types. This is in accord with the fundamental concept that basaltic

Table XXII Chemical Compositions of Ultramafic Rocks (in wt. %)

Oxides	Average Amphibole Peridotite (Nockolds, 1954)	Average Pyroxene Peridotite (Nockolds, 1954)	Average Peridotite (Nockolds, 1954)	Lizard Peridotite (Green, 1964)	Garnet Peridotite Xenolith (Carswell and Dawson, 1970)	Garnet Lherzolite (Boyd and Nixon, 1973)	Pyrolite (Ringwood, 1966)	Average Pyrolite (Ringwood, 1975)	Average Hornblendite (Nockolds, 1954)	Average Eclogite (Hyndman, 1972)	Average Ultramafic Rock (Nockholds, 1954)
SiO_2	43.07	42.05	43.54	44.8	46.5	44.54	45.2	45.1	42.00	47.6	43.8
TiO_2	0.25	0.96	0.81	0.2	0.3	0.26	0.7	0.2	2.86	1.3	1.7
Al_2O_3	4.20	5.50	3.99	4.2	1.80	2.80	3.5	4.6	11.39	14.6	6.1
Fe_2O_3	2.19	2.62	2.51	8.2	6.7	10.24	0.5	0.3	5.27	3.6	4.5
FeO	9.40	9.90	9.84				8.0	7.6	10.30	8.6	8.7
MnO	0.26	—	0.21	0.1	0.1	0.13	0.14	0.1	0.24	0.2	0.18
MgO	36.43	30.79	34.02	39.2	42.0	37.95	37.50	38.1	12.35	8.5	22.10
CaO	2.98	4.78	3.46	2.40	1.5	3.38	3.1	3.1	11.31	10.6	10.10
Na_2O	0.38	0.81	0.56	0.20	0.2	0.43	0.6	0.4	1.80	2.8	0.80
K_2O	0.15	0.44	0.25	0.05	0.2	0.14	0.13	0.02	0.84	0.5	0.70
H_2O	0.64	1.69	0.76	—	—	—	—	—	1.31	—	0.60
P_2O_5	0.05	0.08	0.05	0.01	0.2	—	0.06	0.02	0.33	—	0.30
$Mg/Mg + Fe$	0.74	0.70	0.72	0.82	0.86	0.78	0.80	0.82	0.43	0.40	0.70

Source Materials

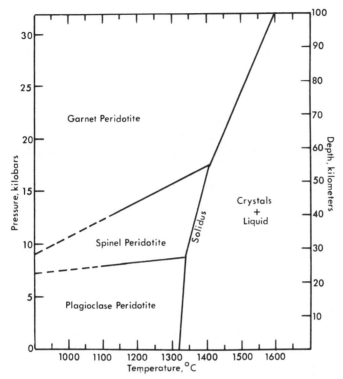

Figure 55 Pressure–temperature stability relations for peridotite compositions. (After Kushiro and Yoder, 1966. With permission of the *Journal of Petrology*.)

magma compositions closely correspond to the invariant points or univariant curves of multicomponent systems (Schairer, 1967).

The seismic velocities (> 8.0 km/sec) and density values (3.3 gm/cc) are in accord with those observed for the mantle (see Table XXIII). Peridotites occur worldwide and possess a narrow compositional range. They are not merely chemical entities but very real petrological substances.

The Mg/Mg+Fe indices are well within the values set for the mantle and peridotites are prominent xenolithic materials.

1.2 Pyrolite Model

Ringwood (1958), following Bowen (1928) and Rubey (1951), suggested a zonal structure for the upper mantle. The complementary association of ultramafics with many basalts led Ringwood (1962) to suggest the upper mantle to be of pyrolite (pyroxene + olivine ± garnet) composition. Such a composition was proposed to lie between peridotites and basalts with possible peridotite/basalt ratios of a high of 4 : 1 and a low of 1 : 1. The recommended composition for pyrolite was essentially a mixture of three parts alpine peridotite (79% olivine + 20% orthopyroxene + 1% spinel) and one part Hawaiian tholeiite. The bulk chemical composition of an average pyrolite is given in Table XXII.

Ringwood and co-workers have emphasized that pyrolite is a *chemical composition*, rather than a mineral or rock composition, with definite ability to form, upon partial melting, basalts with complementary refractory peridotites. This leads to the idea of a widespread chemical and petrological zoning in the upper mantle. Ringwood (1962) and Green and Ringwood (1963) represent the zoning by the following mineral assemblages characterizing pyrolite compositions:

Ampholite	Olivine + amphibole
Plagioclase pyrolite	Olivine + Al–poor pyroxenes + plagioclase
Pyroxene pyrolite	Olivine + Al pyroxenes ± spinel
Garnet pyrolite	Olivine + Al–poor pyroxenes + pyrope-rich garnet

The anhydrous solidus and subsolidus stability relations for the pyrolite compositions have been studied by Green and Ringwood (1967a,b, 1970) for temperature of 800–1500°C and pressure of up to 35 kb. With increasing pressure the subsolidus assemblage changes from (Fig. 56):

plagioclase pyrolite → pyroxene pyrolite → garnet pyrolite

The seismic velocity increase in the upper mantle could be explained by such a transition of pyrolite composition, from a less dense to a more dense product as a function of depth (see Table XXIII). The pressure-dependent changes in pyrolite are qualitatively similar to those observed by Kushiro and Yoder (1966) for peridotite.

Ampholitic compositions may thus be considered to be present in the wedges overlying the Benioff Zone and immediately below the mantle, whereas plagioclase pyrolite may be important in midocean ridges and the island arc system of Japan and the Basin and Range province of western United States. A minor upsurge in seismic velocity 90 km below the central United States may be related to pyroxene pyrolite-garnet pyrolite change (Green and Ringwood, 1970).

Source Materials

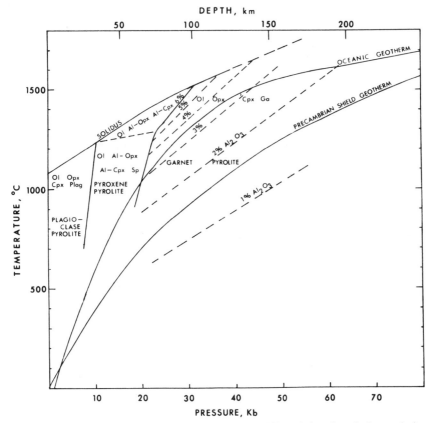

Figure 56 Pressure–temperature solidus and subsolidus stability relations for anhydrous, plagioclase, pyroxene, and garnet pyrolite compositions. Oceanic and shield geotherms indicate stable mineralogy with depth. Ol + Al–Opx + Al–Cpx is favored assemblage at intermediate pressures and high temperatures without any additional aluminous phase. (After Green and Ringwood, 1967. With permission of the *Earth and Planetary Science Letters.*)

Recently Green and Liebermann (1976) presented an integrated model for the upper mantle (cf. Ringwood, 1966) to a depth of 140 km, taking into consideration the data on melting of pyrolite (both anhydrous and with small H_2O and CO_2 contents). This model is shown in Fig. 57 and clearly emphasizes the petrological and the chemical zoning in the mantle as a function of depth with accompanying changes in the densities and stability limits of the assemblages. There is a clear pressure-dependence stability of garnet pyrolite, and it is a major component of the "low velocity zone." The incipient melt portion in

Table XXIII P-Wave Velocity and Density for Selected Compositions

	Composition	Density (gm/cc)	V_P (P-Wave Velocity) (km/sec)
Ultramafic rocks	Dunite[a]	3.32	8.42
	Harzburgite	3.30	8.23
	Peridotite[a]	3.31	8.32
	Garnet peridotite ($Fo_{50}Px_{40}Ga_{10}$)	3.34	8.25
Mean		3.318	8.31
	Eclogite[b]	3.35–3.59	7.65–8.36
Pyrolite[a]	Ampholite	3.27	7.98
	Plagioclase pyrolite	3.26	8.01
	Pyroxene pyrolite	3.33	8.18
	Garnet pyrolite	3.38	8.38
Mean		3.312	8.138
	Upper mantle	3.3–3.6	8.1–8.25
Minerals	Forsterite	3.32	8.44
	Pyroxene	3.28	8.02
	Garnet	3.38	8.08
	Serpentine[c]		5.10

[a] Ringwood, 1966, at 1 atm.
[b] Birch, 1970, at 1 atm.
[c] Christensen, 1966, at 10 kb.

Others calculated from projections of densities on seismic curves.

equilibrium with garnet pyrolite decreases from about 2% at 90–100 km to 0.5% below 100 km depth.

The parent material capable of generating basaltic magmas undoubtedly lies in the upper mantle. It may be garnet peridotitic or pyrolitic (see Table XXII). Pyrolite being a chemical composition may never be truly represented by a corresponding rock but has close similarities to the ones proposed. It makes a chemical basis for petrological reasoning and conclusions.

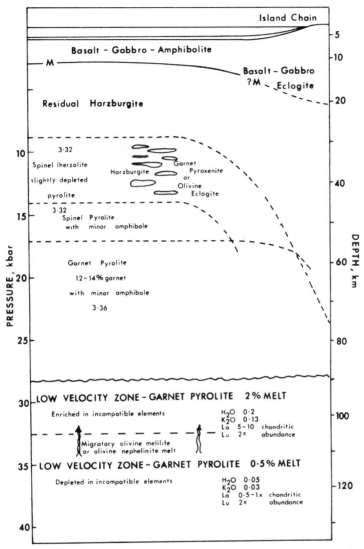

Figure 57 An integrated compositional model for the oceanic lithosphere and the low velocity zone based on high pressure and temperature experimental work on pyrolite and basaltic compositions and physical properties of minerals. (After Green and Liebermann, 1976. With permission of *Tectonophysics*.)

2 MELTING AND MAGMA FORMATION

Having established the nature of the source material, it logically follows that we discuss the process of basaltic magma generation and the compositional characteristics obtained in it. The phase equilibrium and melting relations studies of the silicate and whole rock systems bring out one aspect of fundamental importance that chemical equilibrium of systems is not dependent on the relative proportions of crystalline phases and liquid for a given set of physical conditions. Such data provide important parameters to delineate the physical and chemical limits of the process of magma generation. There are, though, many complexities such as heat of melting, high pressure fractionation of the melt, partitioning behavior of minor elements into crystal/liquid phases, rate of magma ascent, degree and rate of initial melting, depth of magma separation, depth-dependent compositional controls of the melt, thermal regimes of the environment, and other physicochemical conditions attendant at the time of and subsequent to the formation of magma. Some of these aspects will be briefly considered within the scope of the present discussion.

2.1 Heat of Melting

Magma formation by melting is an energy intensive process. Bowen (1928) had suggested that 100 cal/gm could be considered as a realistic heat of melting value for basalt. Yoder (1975) determined heat of melting values, at 1 atm, for a number of compositions corresponding to the invariant points in systems related to basalts, eclogites, and peridotites. These are listed in Table XXIV.

The heat of melting values at 1 atm are lower than the value proposed by Bowen. However, those determined for mantle conditions are relatively much higher (118.5–135.4 cal/gm at 40 kb). Though the heat of melting values are within the thermal energy framework expected in the mantle, they necessitate the availability of large quantities of heat on a continuing basis for beginning and sustaining melting on a volcanogically reasonable scale.

Where does this heat come from? Radioactive heat, principally from the decay of U, Th, and K, seems one probable source. Table XXV lists the abundances of radioactive elements, along with their heat production rates, in crustal and possible mantle materials. There abundances are low, and consequently the

Table XXIV Heat of Melting for Basaltic Compositions

Composition (wt. %)	System	Heat of Melting (ΔH_m) (cal/gm at 1 atm)
$Di_{58}An_{42}$	Eutectic E in the system Di–An (Fig. 1)[a]	77.8
$Di_{49}An_{43.5}Fo_{7.5}$	Piercing point E in the system Di–Fo–An (Fig. 10a)	85.6–79.6
$Di_{28.5}Fo_{4.5}An_{33.5}Ab_{33.5}$	Piercing point on PQ in the system Di–Fo–An_{50}–Ab_{50} (Fig. 11)	73.5
$Di_{34}Py_{66}$	Eutectic in the system Di–Py at 30 kb	83.4 (118.5 at 30 kb)
$Di_{47}Fo_6Py_{47}$	Eutectic E in the system Di–Fo–Py at 40 kb (Fig. 63)	91.4 (134.4 at 40 kb)
$Di_{47}En_3Fo_3Py_{47}$	Invariant point in the system Di–Fo–En–Py	89.5 (135.4 at 40 kb)

[a] Figure numbers refer to the figures in the text. Based on data from Yoder (1975).

Table XXV Abundances of Radioactive Elements and Their Heat Production Rates

Rock Type	Density (gm/cc)	Amount of Radioactive Element (in ppm)[a]			Q (in cal/gm/yr)
		Uranium	Thorium	Potassium	
Granite	2.75	4	13	4	71.7×10^{-7}
Basalt	2.91	0.5	2	1.5	11.96×10^{-7}
Peridotite	3.31	0.02	0.06	0.02	0.24×10^{-7}
Pyrolite II	2.82	0.059	0.25	0.09	1.12×10^{-7}

[a] From Press and Siever (1978) and Ringwood (1958).

heat production is also low and time-dependent.* It may therefore be appropriate to conclude that the heat loss through dissipation, its time-dependent production-accumulation, and somewhat diffused nature does not permit radioactive heat to be exclusively responsible for causing melting. However, it certainly is an important source of heat.

*The rate at which radioactive heat is produced is independent of the temperature and depth but a function of time and the concentration of radioactive elements in the parental materials. If we assume that there is no heat loss and the materials are at the temperature of beginning of melting, then the time required to produce heat equivalent to the heat of melting for a given composition can be expressed by the following relationship:

$$T = \frac{\Delta H_m}{Q}$$

where T = time in years; ΔH_m = heat of melting (enthalpy) of the composition in cal/gm, and Q = amount of heat produced by the composition in cal/gm/yr.

Let us now calculate the time required to produce heat equivalent to melt the eutectic composition $Di_{47}Fo_6Py_{47}$ in the system Di–Fo–Py at 40 kb. Compositionally the system Di–Fo–Py corresponds to a garnet peridotite and eutectic composition to a picritic basalt.

GIVEN ΔH_m = 134.4 cal/gm (see Table XXIV)

$Q = 0.24 \times 10^{-7}$ cal/gm/yr (see Table XXV)

Now the time required will be as follows:

1 Time for complete melting of peridotite

$$T_1 = \frac{\Delta H_m}{Q} = \frac{134.4}{0.24 \times 10^{-7}} = 560 \times 10^7 \text{ or } 5.6 \times 10^9 \text{ years (cf. pyrolite)}$$

2 Time for melting of 5% fraction of peridotite

$$T_2 = T_1 \times \frac{5}{100}$$
$$= 5.6 \times 10^9 \times 0.05 \text{ or } 280 \times 10^6 \text{ years}$$

3 Time for melting 1% fraction of peridotite.

$$T_3 = T_1 \times \frac{1}{100} = 5.6 \times 10^9 \times 0.01 = 56 \times 10^6 \text{ years}$$

Thus to melt, at 140 km depth, 1–5% of basaltic fraction of the peridotite, it will take 56–280 million years. Obviously, if heat losses are taken into account, the time requirement will increase. The volatiles may, however, significantly lower the heat requirements.

Melting and Magma Formation

Magmas may thus form not by complete but by partial melting of the parental materials.

A certain degree of partial melting may be natural at the boundaries of compositional and/or phase changes within the mantle. Also as the melting temperature of minerals increases with pressure (depth) any release effect will induce melting. The relatively large liquidus and solidus intervals ($\sim 200°C$) observed for the melting of peridotitic and pyrolitic compositions at mantle conditions are also supportive of magma formation by partial melting. According to Yoder (1975) adiabatic rise of hot mantle material to shallower depths may also be an important source of heat energy for the melting phenomenon. Frictional heat may be locally important (also see Bailey, 1970).

2.2 Low Velocity Zone

Seismic attenuation data indicate that the "low velocity zone," at a depth of 70–150 km, appears to be the appropriate place for the formation of basaltic magmas. It is believed to be characterized by incipient melting, as shown by the decrease in the velocity of seismic waves (Anderson and Sammis, 1970; Press, 1962; Wang, 1972) at that depth. The estimates, for the degree of incipient or partial melting (or the amount of intergranular liquid) based on 3–10% decrease in the velocity of seismic waves, vary from 4 to 8% (Ringwood, 1975; Yoder, 1976). Only 1% incipient melting, however, could account for the low velocity zone (also see description on pages 177–179).

Lambert and Wyllie (1968, 1970) reported that amphiboles break down at pressures of "low velocity zone." Such a breakdown produces anhydrous phases and free water. The free water may cause a depression (of up to 400°C at 100 km depth) in the melting temperature of anhydrous phases resulting in a small degree of partial melting to produce the "low velocity zone." The solubility of water in silicate melts does increase with depth. The incipient melt may thus hold dissolved water. The availability of carbon dioxide through decarbonation reactions has the potential of further depressing the melting temperature of the mantle materials to increase the chances of partial melting as a natural course. Wyllie and Huang (1976) propose that CO_2 is as good as water in causing incipient melting of mantle peridotite, which must begin to melt in the presence of CO_2 at depth of 80–100 km. The proportions of H_2O and CO_2 will be important for degree of melting. Thus the "low velocity zone" is possibly a result of incipient melting of mantle material. Such incipient melting may be important for mass transfer and large-scale movement of oceanic plates in tectonic processes.

The depth of magma formation may also be dependent on the structural characteristics of a given region. Solomon (1972) proposed that a small amount (0.2%) of interstitial liquid may also exist up to a depth of 400 km to account for elastic characteristics of that region. Sacks and Okada (1974) estimate from the seismic attenuation data, the depth of melting (magma source) for Katmai region, Japan, to be at 300–400 km, whereas it very well may be at 400 km for Chile–Peru region.

2.3 Magma Ascent and Magma Separation

The separation of magma from its locale is dependent on degree of partial melting of the parental material. The estimates of temperature distribution in the earth (Clark and Ringwood, 1964) are not within the limits to cause any significant melting in the anhydrous peridotite.

How then does this mostly solid material dissociate itself and prepare for an ascent? The most favored mechanism seems to be the ascent by "diapiric uprise" of the gravitationally unstable mantle material. This has been advocated by Green and Ringwood (1967a), Ringwood (1969), Wyllie (1971a,b), and Ramberg (1972). The gravitational instability may be caused by density differences in the mantle material or by incipient melting, perhaps 1%, caused by a small amount of water in the "low velocity zone" thus making the material less dense and viscous. Such an incipient melting could begin the very process of magma formation. If the ascent is rapid, then during ascent such a mass probably cools adiabatically [cf. adiabatic rise of Yoder (1975)]. This adiabatic cooling may provide the portion of energy necessary to bring the material well within the beginning of melting range. With ascent the degree of melting may reach 2–30%* (Ringwood, 1969), and the process of magma separation may initiate. The magma separation from an average diapir generally takes place when 20–40% partial melting is obtained. Obviously the degree of partial melting will be higher for rapidly ascending diapirs where solid and melt equilibrium pertains.

Yoder (1975) reported that eclogitic mantle material, rising adiabatically

---

*The eventual melting will be related to the initial depth and rate of the "diapiric uprise" and the extent of solid–liquid equilibrium. Cawthorn (1975) proposes the eventual melting to reach 29 and 69% for diapirs originating at 150 and 300 km, respectively (see Table XXVI) for magmas separating at or near surface. Arndt (1977) suggests 50–80% melting to form ultrabasic (komatiitic) melts.

Melting and Magma Formation

from a depth of 160 km, but maintaining solid + melt equilibrium, will be almost completely molten (basaltic) upon arrival at the surface. Cawthorn's (1975) estimates on percent partial melting in rising diapirs calculated from thermodynamic parameters are given in Table XXVI. However, once the magma separation takes place, the chemical equilibrium with the host is broken. It then follows crystallization and accompanying fractionation independently. The composition of the initial liquid and that which finally separates will be different. Thus the depths of initial melting and final magma separation are probably critical to the magma composition.

Green and Ringwood (1967a) and Wyllie (1971b) have suggested that solid diapirs may also be considered to rise from below the "low velocity zone." Wyllie (1971b) proposed that juvenile water rising upward from deeper levels in the mantle may cause the required gravitational instability and density decrease to begin the ascent. Such an adiabatic rise will cause the material in the diapir to be at higher temperatures than the enclosing material (see levels, *ab, cd,* and *ef* in Fig. 58) thus bringing it in the beginning of melting range. The greater uprise would cause increased melting. This model has implicit suggestion of a certain degree of chemical interchange when such diapirs reach the "low velocity zone."

Yoder has suggested magma ascent by magmafracting "the effective decrease in strength of rocks by virtue of generation of magma and the process of crack propagation by magma in the brittle failure region" (Yoder, 1976, pp. 179–180). Such a process is related to the direction of principal stress.

Table XXVI Depth-Degree of Melting Relationships in Diapirs[a]

Depth	Initial Temp. (°C)	Depth of Magma Separation = 30 km		Depth of Magma Separation = Surface	
		Final Temp. (°C)	% Melt	Final Temp. (°C)	% Melt
150	1760	1412	24	1372	29
225	1950	1545	38	1497	47
300	2200	1647	56	1606	69

[a]From Cawthorn (1975).

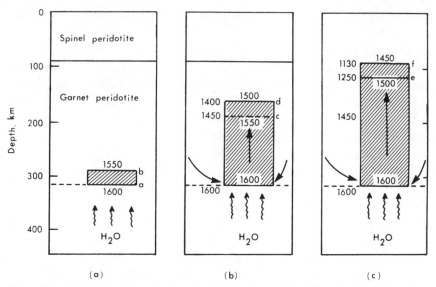

Figure 58 Adiabatic diapiric uprise of a layer *ab;* to *cd* (as in Fig. 58*b*) and *ef* (as in Fig. 58*c*.) The uprise is aided by water migrating from the deep mantle resulting in increased liquid (melt) formation in layer *ab* toward the top of the layer. (After Wyllie, 1971. With permission of John Wiley & Sons, Inc., New York.)

The rates of magma generation and magma rise* will be important factors in magma composition.

2.4 Depth-Degree of Melting and Magma Compositions

The formation of basaltic magma is widely acknowledged to be related to partial melting in the mantle. Eaton and Murata (1960), from their studies of the Hawaiian volcano, reported that magmas can be seismically traced from their point of generation to final eruption. However, the magma compositions will not only be a function of depth-degree of melting but also the compositional characteristics of the parental material, depth of magma separation, and extent of depth-dependent fractionation. In general terms, the magma (basalt) should have equilibrium relationship with the mantle material (peridotite) if it represents a partial

---

*Marsh (1979) estimates an ascent velocity of 10–100 m/yr for calc-alkali magmatism. In the case of magma ascent by elastic-crack propagation, where the length/width ratio *(R)* is 1000, the ascent rate must be at least 30 km/day for magma to reach the surface as all liquid.

melt obtained from it. Let us now briefly examine the depth-composition relations of magmas in the light of data on synthetic and natural compositions.

Kuno (1959), from petrographical, chemical, and field relations, proposed that compositions of basalts may be a reflection of the depth of melting. Silica-poor alkali basalts probably form at a deeper level by partial melting of mantle material (garnet peridotite) than silica-saturated tholeiites. Yoder and Tilley (1962), from their high pressure melting studies of basalts and eclogites, suggested that it is pressure-dependent crystal fractionation of the initial liquid that controls the composition of basaltic magmas. Alkali basalts may result from high pressure fractionation of partial melt in the mantle, whereas its low pressure fractionation yields tholeiites (see Fig. 32). This is apparent from the pressure-dependent shift of phase boundaries toward silica undersaturation schematically assessed from the Fo–Ne–Sil system (see Fig. 47).

Further support to the depth-composition relations of basaltic magmas is provided by O'Hara (1965) from assessment of schematic relations in the general system Cpx–Ol–Ne(Mel)–Sil up to anhydrous pressure of 30 kb. On the basis of shifts, with increasing pressure, in the phase boundaries ("thermal barriers") and composition of the initial liquid (at or near an invariant point), toward the Ol–Ne surface (silica undersaturation), he proposed that the initial liquid from the partial melting of garnet peridotite, found in diamond pipes, will be:

1 Picrite basalt (hypersthene-normative) at high pressure (80–100 km depth).
2 Picrite basalt (nepheline-normative) at intermediate pressure (35–80 km depth).
3 Alkali basalt–high Al-basalt–tholeiite at low pressure (less than 35 km depth).

In turn these may form minor fractionates of phonolites from alkali basalt and rhyolites from tholeiites.

O'Hara and Yoder (1967) reported that initial liquid from the melting of mixtures of Fo + En + Di + Ga (synthetic garnet peridotite) in the system $CaO–MgO–Al_2O_3–SiO_2$ at 30 kb is indeed hypersthene-normative picrite basalt type. O'Hara (1968) stressed that closed system high pressure continuous olivine fractionation (20–40%) from picrites, during ascent, is important to the derivation of a variety of surface basalt magmas. Lack of olivine as a liquidus phase in olivine tholeiites (see Fig. 49) at a pressure of 12 kb or more is interesting in this regard, as it suggests nonequilibrium relations of olivine tholeiite and peridotite.

The occurrence of picritic lavas and inclusions of garnet harzburgites, eclogites, and lherzolites in alkali basalts is suggested as supportive evidence to a degree of fractional crystallization of the initial liquid at depth. O'Hara's high pressure magma separation and olivine fractionation scheme is in direct contrast to the low pressure separation and fractionation of mafic magma model of Green and Ringwood (1967a). Ringwood (1975) acknowledges the role of low pressure olivine fractionation, but he objects to O'Hara's scheme, as 20–40% olivine removal from picrites would allow lower Ni and Cr contents than those found in inclusion-bearing alkali basalts and tholeiites. Also the Mg/Mg + Fe ratios would be 0.55 or less rather it is found to be between 0.63 and 0.73.

Kushiro (1968) observed that the Fo–En boundary and piercing point in the system Fo–Ne–Sil shift toward the nepheline apex (silica undersaturation; Figs. 47 and 59) with pressure increasing to 30 kb. He proposed the following scheme for the composition of the initial liquid:

1 Below 10 kb (35 km depth), quartz-normative basalt
2 Between 10–30 kb (35–105 km depth), olivine-hypersthene-normative basalt
3 At 30 kb (105 km depth), nepheline-normative basalt

thus implying depth-composition dependency of basalts. Alkali basalts therefore form at greater depths than tholeiites.

Eggler (1974) found that the shift in the liquid composition toward silica undersaturation persists in the system Fo–Ne–Sil at pressures of 20 kb in the presence of water and carbon dioxide (Fig. 47).

Green (1973a,b) integrated the melting data on basaltic and pyrolitic compositions under different pressure–temperature and volatile pressure conditions, and the knowledge of the nature of the solid–liquid phases into a petrogenetic grid for basalt formation, both under the oceans and island arc systems. The grid is shown for water-undersaturated conditions in Fig. 60 and for water-saturated conditions in Fig. 61. The diagrams also show the compositional characteristics of the parent material, degree of partial melting, and probable depths of magma separation for various depths of partial melting.

According to this model, the direct partial melting (10–15% with magma separation at a depth of 20 km under oceans, Fig. 60; 20–30% with magma separation at a depth of 40–60 km under island arcs, Fig. 61) is capable of producing quartz-normative basaltic magmas. Water-saturated conditions in an island arc system will favor calc-alkaline tendencies.

Melting and Magma Formation

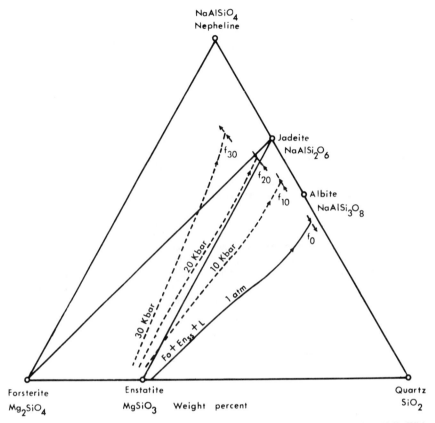

Figure 59 The system forsterite–nepheline–silica at various pressures. (After Kushiro, 1968. With permission of the *Journal of Geophysical Research*.) Note the composition of the piercing point and forsterite–enstatite phase boundary move toward the nepheline apex (silica undersaturation) with increasing confining pressure. Dashed lines are estimated boundaries (1 atm data from Schairer and Yoder, 1961).

The production of calc-alkaline magmas by direct partial melting of the hydrous mantle has also been proposed by Yoder (1969). Kushiro (1970) reported that initial liquids from partial melting of garnet peridotite under water-saturated conditions at 20–25 kb tend to be quartz-normative basalts or even andesitic basalts. The liquid composition is nephelinitic above 30 kb water-undersaturated conditions (also see Fig. 47).

Further, the model proposes that a very small degree (2–5%) of partial melting of parental mantle material is required to produce highly alkaline and

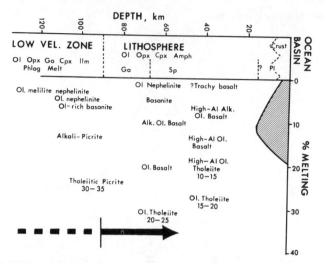

Figure 60 Integrated petrogenetic grid (after Green, 1973) based on melting studies of pyrolite and basaltic compositions showing the relationship of depth of magma separation-degree of partial melting to magma compositions in the oceanic crust-mantle situation for water-undersaturated conditions of melting of pyrolite. The mineralogical character of the lithosphere and depth of onset of melting is given at the top. The hatched area is the zone of the derivation of quartz-normative melts by direct partial melting of the mantle. (With permission of the *Earth and Planetary Science Letters*.)

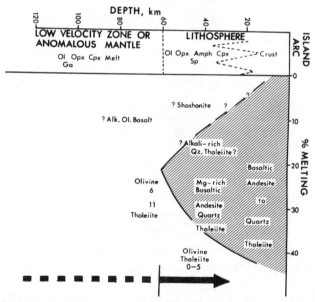

Figure 61 Petrogenetic grid (after Green, 1973) showing the relationship of depth of magma separation-degree of melting to magma composition under island-arc situation for water-saturated conditions. The mineralogical character of the lithosphere and depth of onset of melting is given at the top. The hatched area indicates derivation of quartz normative magmas by direct partial melting of the mantle. (With permission of the *Earth and Planetary Science Letters*.)

Melting and Magma Formation

silica-undersaturated basaltic melts under the oceans. The silica-undersaturation and alkaline character (olivine nephelinitic-melilititic) increases with increasing depth of initial melting and magma separation (60 km). However, for a given depth of magma separation the increase in degree of melting (2–30%) forms relatively silica-rich basaltic magmas.

Magmas produced and separated below 90 km are generally much more alkaline- and silica-undersaturated than those at shallower depths.

Ringwood (1975) proposed a detailed and comprehensive model for magma line of descent as a function of temperature, depth, and degree of partial melting of pyrolite mantle containing 0.1% water. This petrogenetic scheme is shown in Fig. 62 and assumes a rising diapir magma model. The magma generation

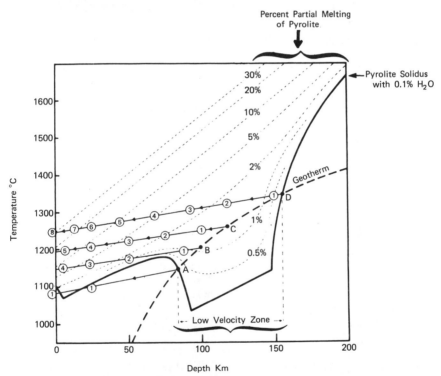

Figure 62 Relationships between degree of melting-depth of magma separation to magma composition in the pressure–temperature field of the mantle solidus for rising diapir situation. A, B, C, and D are positions of diapirs. Lines from them show the pressure–temperature paths of rising diapirs. Numerals represent magma compositions at depths of separation for various degrees of partial melting. Solid curve is the mantle solidus for pyrolite composition with 0.1 wt. % H_2O. (After Ringwood, 1975. With permission of McGraw-Hill, Inc., New York.)

is believed to occur in the "low velocity zone" (70–150 km depth), where the beginning of melting (solidus) curve for pyrolite intersects the geothermal gradient, thus making it a thermally suitable place for the initiation of melting. The rising diapirs in Fig. 62 are denoted by A, B, C, and D. The numbers 1, 2, 3, 4, . . . represent the depths of magma separation at that degree of partial melting. The detailed compositional scheme for various depths of magma separation and degrees of partial melting is given in Table XXVII.

It is apparent from this model that the degree of eventual melting increases with increasing depth of initial melting and the extent of solid and liquid equi-

Table XXVII Magma Generation and Low Pressure Liquid Lines of Descent

Depth of Magma Separation	Diapir in Fig. 62		Magma Types	Partial Melting (%)
70 km	A		High temperature peridotite	<1
100 km ↓ 50 km ↓ Surface	B	1 2 3 4	Olivine nephelinite Olivine basanite High Al-alkali basalt Quartz tholeiite	1 ↓ 8
120 km ↓ 50 km ↓ Surface	C	1 2 3 4 5	Olivine nephelinite Olivine basanite Alkali olivine basalt High Al-basalt Quartz tholeiite	1 ↓ 15
150 km ↓ 45 km ↓ Surface	D	1 2 3 4 5 6 7 8	Kimberlite Olivine nephelinite Olivine basanite Alkali olivine basalt Olivine basalt High Al-olivine tholeiite Olivine tholeiite Tholeiitic picrite	0.5 ↓ 5 ↓ 18 ↓ 30

Melting and Magma Formation

librium. The alkaline- and silica-undersaturation character of the initial liquid becomes dominant with depth. However, the depth of magma separation and degree of partial melting exert dominant control on eventual magma compositions. The general scheme may be summarized as follows:

1 Near surface magma separation—tholeiitic.
2 Separation near 50 km—alkali olivine basaltic–basanitic.
3 Separation near 100 km or more—olivine nephelinitic–kimberlitic.

Though the above models differ in some details, the overall depth-composition relations of magmas are apparent. In the lack of direct observation of depth of magma separation, the above models may be appropriate to a degree or for specific cases. The detailed examination of the major and trace element characteristics of basalts may provide further supporting evidence for the models. The new understanding about the role of volatiles in mantle does, however, suggest the possibility of magma formation at lower pressures and temperatures than previously proposed.

2.5 Compositional Consistencies in Voluminous Lava Flows

Now let us examine the question as to what type of melting process is responsible for bringing to the surface enormous volumes of consistently homogeneous composition of basalt flows as represented by the world's ancient volcanic plateaus. It is evident from Table XXVIII that a good rate of lava extrusion was responsible for thick accumulations of lava flows of the major volcanic plateaus. The midoceanic ridge system is a good present-day example of accumulation of lava at a substantial rate. Yoder (1976) suggested that the compositional consistency of the basaltic flows may be explained by isothermal withdrawal of liquids when batch (partial) melting of such a parent is considered to represent eutectic composition. The volume of flows being controlled by degree of partial melting.

Yoder used the phase relations in the system Di–Fo–Py at 40 kb, studied by Davis and Schairer (1965), as a basis of this explanation. The end members of the system are important components of garnet peridotite. The system was considered to be a "ternary" with piercing point E to be a eutectic at 1670°C (Fig. 63). The temperature of E is close to the temperature of beginning of melting of compositions in the system and that estimated at 140 km depth. The com-

Table XXVIII Rates of Lava Extrusions in Flood Basalts and Midoceanic Ridges[a]

Locality	Volume (km^3)	Rate of Extrusion/Year (km^3)
Columbia River Plateau, USA	195,000	0.02
Deccan Plateau, India	500,000	0.025
Siberian Plateau, USSR	900,000	0.0069
Karoo lavas, South Africa	1,400,000	0.014
Parana Plateau, South America	780,000	0.026
Triassic Nikolai flows,[b] Alaska; British Columbia	300,000	0.06
Midocean ridges	—	5.0
World volcanics 490 volcanoes	—	2.0

[a] Based on data in Yoder (1976).
[b] Estimated from available data.

position of the eutectic ($Di_{47}Fo_6Py_{47}$; Fig. 63) has normative similarities to basalt (picrite).

1 A composition such as X (Fig. 63), begins to melt at the eutectic temperature to a liquid of eutectic composition E. As the eutectic is attained, the liquid proportion, according to the Lever Rule, will be 30%. The residual unmelted fraction will be represented by a point R on the garnet–forsterite join (Fig. 63). If such liquid is isothermally withdrawn in a batch, it will have eutectic (or similar) composition throughout its withdrawal. This may be useful to explain the compositional consistency of basalt flows. The volume of the flow being related to the degree of melting.

2 If the temperature is further raised, before the withdrawal of the first liquid, the liquid composition will then move to point A (Fig. 63) on the Fo–Py phase boundary, but directly away from the forsterite apex through the composition X assuming all diopside is melted. The liquid fraction at point A will be 46% and all the pyrope is also melted. At this stage, the liquid leaves

the boundary along the line A–Fo (Fig. 63) and complete melting occurs at the temperature of point X. This consideration provides a mechanism by which a series of liquids between the initial composition of E and the final composition of X can be produced. At any stage of removal, the liquids will have an olivine/hypersthene normative composition.

3 Consider now the case when only a few percent of liquid E is removed. Removal of liquid E (Fig. 63) as it is formed is an example of fractional melting. A series of such withdrawals will move the residual compositions from $X \rightarrow X' \rightarrow X''$ (Fig. 63) but the composition of the liquids produced will be similar (a method of obtaining periodic magma pulses).

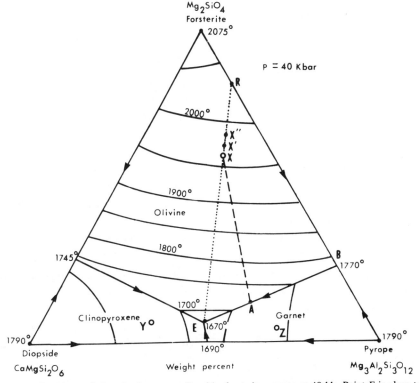

Figure 63 Phase relations in the system diopside–forsterite–pyrope at 40 kb. Point E is close to being a eutectic. The system can be considered a compositional plane (after Davis and Schairer, 1965; with permission of the Carnegie Institution of Washington). Crystallization paths are from Yoder, 1976 (with permission of the National Academy of Sciences).

4   Assuming compositional heterogeneity in the mantle, consider mantle compositions to be those represented by the points X, Y, or Z (Fig. 63) each having different proportions of forsterite, diopside, and pyrope. In each case the first liquid formed upon melting will have the same composition, that is, that of point E. However, the liquid compositions obtained after the eutectic is attained will be different in each case.

The above cases, though highly simplified, serve to illustrate the mechanism of obtaining compositionally consistent basalts. The extent of solid solution and proportion of phases being consumed will obviously influence the minor element contents of the basalts. Also, the eutectic liquids in this system (or the primary liquids obtained from partial melting of garnet peridotite) will be picritic and would require olivine (or Ol + En) fractionation to give basalts (Davis and Schairer, 1965), which is in conformity with proposals of O'Hara (1965, 1968), O'Hara and Yoder (1967), Ito and Kennedy (1967), and Yoder (1974).

2.6   Role of Volatiles

The role of water in melting processes in the mantle to form magmas was discussed earlier. However, since Eggler (1973) showed that large amounts of CO_2 can dissolve in silicate melts, the role of CO_2 in magma generation has gained considerable attention. Eggler (1974) has greatly stressed the role of CO_2 in the generation of alkaline magmas, particularly the silica-undersaturated melts. A small amount of CO_2 (0.15–0.35%) in the mantle can cause the formation of alkali basaltic magmas at a pressure 7 kb lower than in the case of an anhydrous condition (also see Eggler, 1978).

Wyllie (1977a) and Wyllie and Huang (1975, 1976) have greatly emphasized that the formation of carbonatitic–kimberlitic and related liquids may be a function of depth and $CO_2(+ H_2O)$ controlled degree of melting. They propose that partial melting of peridotitic mantle, in the presence of CO_2, at depths less than 80 km, will produce basaltic magmas. At depths near 100 km, the melt becomes progressively silica-deficient so that the first small quantities of liquids produced are essentially carbonatitic. With further melting they change to kimberlites and finally to basalts, but at much higher temperatures. The projected solubility of large amounts of H_2O (up to 34 wt. %; Brey and Green, 1977) and CO_2 (up to 40 wt. %; Wyllie and Huang, 1976) in silicate melts under mantle conditions, and consequent depression of melting temperatures by ~400°C bring the thermal regimes of magma formation within the realities of natural geothermal conditions prevalent in the upper mantle.

3 SUMMARY

It is agreed that the source material for basaltic magma is garnet peridotitic/pyrolitic mantle material. Basaltic magma most probably results from the following:

1. Partial melting of peridotite or pyrolite mantle material in the pressure–temperature conditions of the "low velocity zone." Radioactive heat, adiabatic uprise, phase-compositional changes, and traces of volatiles may collectively effect incipient melting in the mantle.
2. The incipient melting is important to "diapiric uprise" of the mantle mass. Melting increases with ascent. At 20–40% melting, magma separation occurs.
3. The degree of eventual melting of the magma is a function of depth of initial melting and rate of ascent. The greater the depth of initial melting, the higher is the degree of eventual melting at the time of magma separation if adiabatic conditions prevail. The final melt fraction is greater for cases of rapid rise if the solid and liquid equilibrium obtains throughout the rise.
4. The depths of initial melting and magma separation are important controls of magma compositions, along with the compositional characteristics of the parental materials.
5. The initial melt is basaltic whose silica undersaturation and alkaline character is a function of depth and/or degree of melting. Only a small degree of melting is required to produce highly silica-undersaturated magmas.
6. The compositional consistency of flood and other basalts may be explained by a eutectic-type batch melting of compositionally heterogeneous mantle material.
7. Once the magma separation occurs, then magma follows independent cooling, crystallization, and fractionation paths. Alkali basalts generally are produced at greater depths than tholeiites or calc-alkaline magmas.
8. Basalt magmas probably undergo some degree of continuous fractionation with ascent.
9. The role of volatiles (H_2O and CO_2) is important in bringing the melting temperatures of mantle materials within the thermal regimes prevalent in the mantle.

CHAPTER 8
Conclusions

It should be apparent, at this point, that the experimental data on petrologically pertinent synthetic silicate and whole rock systems, under different pressure and temperature conditions, are extremely important and useful in formulating the physiochemical framework of the igneous process from magma generation to its final consolidation. Figure 64 schematically displays major stages of the igneous process.

It may be concluded, from phase equilibrium studies at low pressures on synthetic silicate systems composed of early and late crystallizing minerals, that the low melting compositions are those rich in silica and/or feldspathic components. Thus fractional crystallization provides a chemical link between the primary and derivative magmas. The presence of large phase areas (volumes) of mafic components in multicomponent silicate systems indicate their dominant and early crystallization even from compositions containing only about 2.0 wt. % of mafic minerals. This is consistent with petrographic and field observations on the order of crystallization of minerals from magma.

The determination of the univariant and invariant characteristics of different systems facilitates developing of flow sheets to depict interrelationships of possible liquid trends with their natural analogs. Such data led to an important conclusion that the bulk chemistry of the initial melt is one of the major controls dictating the compatibility and incompatibility of minerals during magmatic crystallization. Thus certain compositional planes, for example, the planes Di–Fo–Ab, Di–Fo–Ne, Di–Ab–Wo in basaltic, and Ab–Or in felsic systems act as thermal divides causing such relationships. Therefore very slight differences in the chemical compositions of the original melt can result, upon crystallization, in distinctly different derivatives. It is possible that such compositional thermal divides may also exist in nature, at least at low pressures, to cause the spatial separation of chemically distinct (silica-undersaturated and silica-oversaturated) rock groups.

It further implies that no single magma, under conditions of equilibrium crystallization, can form both silica-undersaturated and silica-oversaturated

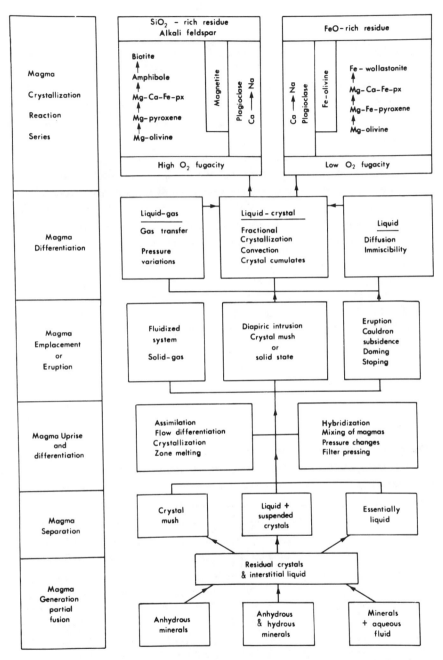

Figure 64 Schematic representation of the major stages of the total igneous process from magma generation to final crystallization. (After Wyllie, 1971. With permission of John Wiley & Sons, Inc., New York.)

residual liquid trends at the same time. However, the factors of solid solution, solid–liquid reaction, pressure, and compositional changes through substitutions would tend to weaken the nature of such thermal divides.

It appears likely that silica-undersaturated magma lineages with both sodic and potassic affinities are connected, by early crystal–liquid controls, to a single composition.

Phase equilibrium data permit another important observation that, irrespective of the initial compositions, melts in the multicomponent systems tend, with a small drop in temperature after the crystallization of the primary phase, to approach a four-phase curve. Such is also the case of volcanic rocks, which, despite poor crystallinity, generally have the major phases present. It is therefore possible that magma compositions lie close to the four-phase curves of the synthetic systems.

Partial pressure of oxygen is an important factor in the igneous crystallization process. In the light of data at hand it can be established that low P_{O_2} favors a crystallization trend toward iron-enrichment (e.g., layered intrusions), whereas high P_{O_2} leads to a silica-enrichment trend (orogenic calc-alkaline rocks).

In contrast to anhydrous conditions, volatile (water vapor) pressure has a remarkable effect on the lowering (400°C or more) of melting and crystallization temperatures. In addition, it has a profound effect on shifting the composition and characteristics of the invariant points (see Di–An, Ne–Ks–Sil and Ab–Or–Sil at 1 atm and 5 kb/P_{H_2O}) and phase boundaries, changing the appearance of phases at the liquidus, and reversed or delayed order of crystallization. It also controls the stability fields of minerals, for example, with increasing water vapor pressure, leucite becomes unstable whereas amphibole is stable in relation to other mafic minerals. These observations explain many of the petrologic irregularities, for example, compositional zoning in minerals, absence of leucite in plutonic rocks, and so forth.

In general, it can be said that increase in pressure shifts the composition of the invariant points toward silica undersaturation. This has important bearing on the depth-composition relationship of magma.

The solubility of water in silicate melts seemingly increases with pressure (depth). Recent measurements indicate that as much as 40 wt. % $H_2O + CO_2$ may be soluble in mafic silicate melts under upper mantle conditions, thus lowering the beginning of melting by almost 400°C. This has tremendous implications on the regimes of magma generation.

The appearance of two liquid fields in many silicate systems may be indicative of liquid immiscibility as an inherent property of magma. Liquid immiscibility

Conclusions

may be aided by volatiles and alkali contents of the melt. It may be a viable mechanism for the formation of selected chemically contrasting, but juxtaposed, for example, interlayered basalts-rhyolites or alkaline-carbonatitic, rocks.

The studies on the melting relations of rocks under hydrous and anhydrous conditions bring out certain points of petrological importance.

Such data show that basalts readily change to amphibolites even under low water vapor pressure. Amphibolites are thus stable equivalents of basalts at moderate pressures. Amphibole becomes a liquidus phase in basalts at 11 kb/P_{H_2O}. All basalts, at appropriate pressure-temperature conditions, change to eclogites.

Under anhydrous pressure conditions, the liquidus phase in basalts changes from olivine (below 10 kb) to orthopyroxene and clinopyroxene (between 10 and 25 kb) and finally to garnet (at 25 kb or more). This has implications on the pressure-dependent fractionation sequences of basalts controlled by pyroxenes at intermediate pressures and olivine at low pressures. There seems to be a reliable relationship between the liquidus temperatures and the iron-enrichment index (FeO + Fe_2O_3/MgO + FeO + Fe_2O_3) values of greater than 0.5 for mafic rocks. The melting temperatures decrease with the increasing index.

Melting relations of silica-undersaturated felsic alkaline rocks show the dependence of melting intervals on the volatile and Na + K/Al ratio of the rocks. The beginning of melting temperatures for rocks with high volatile contents are near 600°C. Therefore, silica undersaturated peralkaline magmas probably have volatiles dissolved in the melt. Fixation of volatiles in the minerals during the crystallization process is responsible for their diverse mineralogy and low consolidation temperatures.

One of the major achievements of the experimental studies on the synthetic and whole rock systems is the recognition of the fact that magma compositions closely approximate the invariant points or univariant lines of the multicomponent systems. Therefore it is suggestive that magmas may themselves be products of melting of more primitive compositions. The pressure-temperature framework of laboratory studies has established that basaltic magmas are the result of partial melting of garnet peridotitic or pyrolitic upper mantle composition in the "low velocity zone." The depth-degree of melting and depth of magma separation are major controls of basaltic magma composition. In general, shallow depth melting and separation produces silica-saturated (tholeiitic) basalts, whereas melting at greater depths forms silica-undersaturated olivine basalt-type magmas. Approximately 20–40% melting is essential to magma separation. Highly silica-undersaturated alkaline basalts (nephelinites-melilite basalts) may be produced through a rather small degree of melting but at greater depths. Thus the continuum observed in basalts may be a function of depth and the consistency in basaltic composition in space and time a reflection of liquids derived at or near an "invariant point" of a multicomponent system chemically analogous to upper mantle composition.

The enormous experimental data on silicate systems and melting relations of basaltic–pyrolitic–garnet peridotitic compositions under mantle pressure–temperature conditions permit the development of a useful and comprehensive depth–temperature–composition framework to 400 km depth. It seems certain that various crustal and surface phenomena are linked to the processes operating in the mantle. Therefore magmatism–plate tectonics–mantle differentiation and crustal evolution may all be manifestations of the vertical and lateral mass transfer of the mantle material, probably and mostly, in the "low velocity zone" (Fig. 65).

Ringwood (1969), in presenting an integrated view of the mantle differentiation, proposed that gravitational instability causes pyrolitic material in the "low velocity zone," but directly beneath the midoceanic ridges, to rise upward. Such continued uprise causes increased melting of the pyrolite to generate basaltic magma and a residual refractory peridotite-type material. More material feeds into this zone *laterally* rather than a *vertical influx* from levels deeper than the "low velocity zone" (see Fig. 65).

As this process of ridge formation is going on, mantle differentation also continues so as to form the following zoned structure (70 km thick).

$$\left.\begin{array}{l}\text{Basalts}\\ \text{Residual peridotites, etc.}\\ \text{Pyrolite}\\ \text{Primitive pyrolite}\end{array}\right\}\text{Oceanic lithosphere}$$

This zone behaves as a single unit and is referred to as the Oceanic Lithospheric Plate. It moves and spreads laterally toward a trench. The trench, a subduction zone at oceanic and continental plate convergence, may initially be formed by gravitational instability resulting from density differences due to phase changes (mafic material → denser eclogite or amphibolite). The descending lithosphere may also cause continued deepening of the trench by itself undergoing phase changes. Partial melting in this sinking plate may form calc-alkaline magmatism characteristic of the island arcs (see Fig. 65) and contribute to the crustal growth. Such partial melting will leave behind a denser residual eclogite

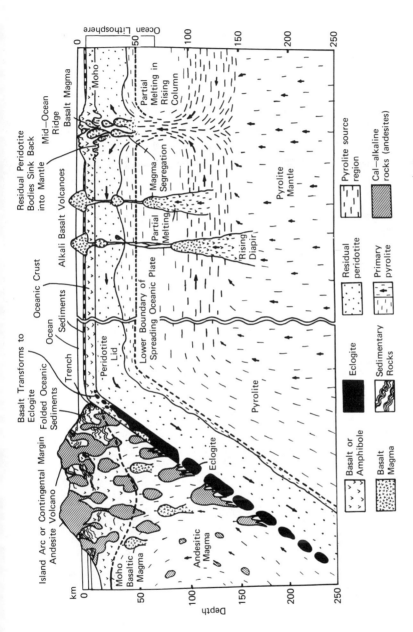

Figure 65 Magmatism, plate tectonics, mantle differentiation, and crustal evolution model. Note the spreading lithospheric plate subducting under the trench to produce calc-alkali magmatism of island arcs. Basaltic magma formation at the ridges is by partial melting in the rising diapirs. The alkali basalts represent diapirs from greater depths. (After Ringwood, 1969. With permission of the American Geophysical Union.)

203

(cf. Green and Ringwood, 1966, dry and wet melting of eclogites), which will continue to sink till the density is balanced by depth or phase changes. Parallel to this process may also be partial melting of the material in the "wedge" (see Fig. 65) to produce basaltic magmas, which add to the continent.

Ringwood (1975) suggested that the sinking differentiated lithosphere continues to sink to depths of 700 km or more. However, in this process the pyrolite portion becomes detached in the upper mantle from the second cycle of the melting process. The heavier portion, peridotite and eclogite, is lost into the deep mantle and is not a part of the later cycles. Such a process thus characterizes "irreversible mantle differentiation," and may be responsible for magmatism–plate tectonics processes in a very fundamental way. Therefore, vertical, rather than lateral, differentiation may have been important in the earliest history of crustal evolution.

References

Akella, J., and F. R. Boyd (1974) Petrogenetic grid for garnet peridotites, *Carnegie Inst. Washington Yearb.*, **73**, 269–273.

Anderson, D. L., and C. Sammis (1970) Partial melting in the upper mantle, *Phys. Earth Planet. Inter.*, **3**, 41–45.

Andersen, O. (1915) The system anorthite–forsterite–silica, *Am. J. Sci.*, **39**, 407–454.

Arculus, R. J. (1974) Solid solution characteristics of spinels: Pleonstate–chromite–magnetite compositions in some island arc basalts, *Carnegie Inst. Washington Yearb.*, **73**, 322–327.

Arculus, R. J., and E. F. Osborn (1975) Phase relations in the system MgO–iron oxide–Cr_2O_3–SiO_2, *Carnegie Inst. Washington Yearb.*, **74**, 507–512.

Arculus, R. J., M. E. Gillberg, and E. F. Osborn (1974) The system MgO–iron oxide–Cr_2O_3–SiO_2: Phase relations among olivine, pyroxene, silica and spinel in air at 1 atm., *Carnegie Inst. Washington Yearb.*, **73**, 317–322.

Arndt, N. T. (1976) Melting relations of ultramafic lavas (komatiites) at 1 atm. and high pressure, *Carnegie Inst. Washington Yearb.*, **75**, 555–562.

Arndt, N. T. (1977) Ultrabasic magmas and high degree of melting of the mantle, *Contrib. Mineral. Petrol.*, **64**, 205–221.

Asklund, B. (1949) Apatitjarnmalmernas differentiation, *Geol. Foeren. Stockholm Foerho.*, **71**, 127–176.

Bailey, D. K. (1966) Carbonatite volcanoes and shallow intrusions in Zambia, in *Carbonatites*, O. F. Tuttle and J. Gittins, Eds., Wiley, New York.

Bailey, D. K. (1970) Volatile flux, heat focussing and the generation of magma, in *Mechanism of Igneous Intrusion*, G. Newall and N. Rast, Eds., Gallery Press, Liverpool, pp. 177–186.

Bailey, D. K. (1974a) Continental rifting and alkaline magmatism, in *The Alkaline Rocks*, H. Sorensen, Ed., Wiley, New York, pp. 148–159.

Bailey, D. K. (1974b) Nephelinites and ijolites, in *The Alkaline Rocks*, H. Sorensen, Ed., Wiley, New York, pp. 53–66.

Bailey, D. K., and J. F. Schairer (1964) Feldspar-liquid equilibria in peralkaline liquids—The orthoclase effect, *Am. J. Sci.*, **262**, 1198–1206.

Bailey, D. K., and J. F. Schairer (1966) The system Na_2O–Al_2O_3–Fe_2O_3–SiO_2 at 1 atm. and the petrogenesis of alkaline rocks, *J. Petrol.*, **7**, 114–170.

Barker, D. S. (1965) Alkalic rocks at Litchfield, Maine, *J. Petrol.*, **6**, 1–27.

Barth, T. F. W. (1948) Recent contributions to the granite problem, *J. Geol.*, **56**, 235–240.

Barth, T. F. W. (1962) *Theoretical Petrology*, 2nd ed., Wiley, New York.

Bell, K., and J. L. Powell (1969) Strontium isotopic studies of alkalic rocks: The potassium rich lavas of Birunga and Toro-Ankole region, East and Central Equatorial Africa, *J. Petrol.*, **10**, 536–572.

Billibin, Y. U. A. (1939) Dissociation of Molecules in the Magmatic melt as a factor in magma differentiation, *Dokl. Akad. Nauk. SSR*, **24**, 783–785.

Birch, F. (1970) Interpretation of low velocity zone, *Phys. Earth Planet. Inter.*, **3**, 178–181.

Bockris, J. O'M., J. L. White, and J. D. MacKenzie (1959) *Physiochemical Measurements at High Temperatures*, Butterworth, London, 394 pp.

Boetcher, A. L., B. O. Mysen, and P. J. Moderski (1975) Phase relations in natural and synthetic peridotites–H_2O and peridotite–H_2O–CO_2 systems at high pressure, *Phys. Chem. Earth*, **9**, 855–867.

Borley, G. D. (1967) Potash-rich volcanic rocks from southern Spain, *Mineral. Mag.*, **36**, 364–379.

Bowen, N. L. (1912) The order of crystallization in igneous rocks, *J. Geol.*, **20**, 457–468.

Bowen, N. L. (1913a) The melting phenomenon of plagioclase feldspars, *Am. J. Sci.*, **35**, 577–599.

Bowen, N. L. (1913b) The order of crystallization in igneous rocks, *J. Geol.*, **21**, 399–401.

Bowen, N. L. (1914) The ternary system diopside–forsterite–silica, *Am. J. Sci.*, **38**, 207–264.

Bowen, N. L. (1915a) The crystallization of haplobasaltic, haplodioritic and related magmas, *Am. J. Sci.*, **40**, 161–185.

Bowen, N. L. (1915b) Crystallization-differentiation in silicate liquids, *Am. J. Sci.*, **39**, 175–191.

Bowen, N. L. (1915c) The later stages of the evolution of igneous rocks, *J. Geol.*, **23**, suppl., 1–89.

Bowen, N. L. (1922) The reaction principle in petrogenesis, *J. Geol.*, **30**, 513–570.

Bowen, N. L. (1928) *Evolution of Igneous Rocks*, Princeton University Press, Princeton, N.J., 332 pp.

Bowen, N. L. (1935) The igneous rocks in the light of high temperature research, *Sci. Mon.*, **40**, 487–503.

Bowen, N. L. (1937) Recent high-temperature research on silicates and its significance in igneous geology, *Am. J. Sci.*, **33**, 1–21.

Bowen, N. L. (1945) Phase equilibria bearing on the origin and differentiation of alkaline rocks, *Am. J. Sci.*, **243-A**, 75–89.

Bowen, N. L. (1947) Magmas, *Bull. Geol. Soc. Am.*, **58** 263–380.

Bowen, N. L., and O. Andersen (1914) The binary system MgO–SiO_2, *Am. J. Sci.*, **37**, 487–500.

Bowen, N. L., and J. F. Schairer (1935) The system MgO–FeO–SiO_2, *Am. J. Sci.*, **29**, 151–217.

Bowen, N. L., J. F. Schairer, and E. Posnjak (1933) The system CaO–FeO–SiO_2, *Am. J. Sci.*, **26**, 273–297.

Bowen, N. L., and O. F. Tuttle (1950) The system $NaAlSi_3O_8$–$KALSi_3O_8$–H_2O, *J. Geol.*, **58**, 489–511.

Boyd, F. R. (1973) A pyroxene geotherm, *Geochim. Cosmochim. Acta*, **37**, 2533–2546.

Boyd, F. R. (1974) Ultramafic nodules from the Frank Smith kimberlite pipe, South Africa, *Carnegie Inst. Washington Yearb.*, **73**, 282–285.

References

Boyd, F. R., and P. H. Nixon (1973) Structure of the upper mantle beneath Lesotho, *Carnegie Inst. Washington Yearb.*, **72**, 431–445.

Boyd, F. R., and P. H. Nixon (1976) Ultramafic nodules from the Kimberley pipes, South Africa, *Carnegie Inst. Washington Yearb.*, **75**, 544–546.

Brey, G. (1976) CO_2 Solubility and solubility mechanisms in silicate melts at high pressures, *Contrib. Mineral. Petrol.*, **57**, 215–221.

Brey, G., and D. H. Green (1975) The role of CO_2 in the genesis of olivine melilitite, *Contrib. Mineral. Petrol.*, **49**, 93–103.

Brey, G., and D. H. Green (1976) Solubility of CO_2 in olivine melilitite and role of CO_2 in earth's upper mantle, *Contrib. Mineral. Petrol.*, **55**, 217–230.

Brey, G., and D. H. Green (1977) Systematic study of liquidus phase relations in olivine melilitite + H_2O + CO_2 at high pressures and petrogenesis of an olivine melilitite magma, *Contrib. Mineral. Petrol.*, **61**, 141–162.

Brotzu, P., L. Morbidelli, and G. Trâversa (1973) Lava e prodotti scoriacei del settore di monterado (bolsena-vulsini orientali) Studio geo-petrografico, *Boll. Serv. Geol. Ital.*, **94**, 1–54.

Brown, G. M., and J. F. Schairer (1967) Melting relations of some calcalkaline volcanic rocks, *Carnegie Inst. Washington Yearb.*, **66**, 460–467.

Bullard, F. M. (1947) Studies on Paricutin Volcano, Mexico, *Bull. Geol. Soc. Am.*, **58**, 433–450.

Burnham, C.W. (1967) Hydrothermal fluids at the magmatic stage. In H. L. Barnes (ed) *Geochemistry of Hydrothermal Ore Deposits*, Holt, New York.

Burnham, C. W., and R. H. Jahns (1962) A method for determining solubility of water in silicate melts, *Am. J. Sci.*, **260**, 721–745.

Cameron, E. N. (1963) Structure and rock sequences of the critical zone of the Eastern Bushveld Complex, *Min. Soc. Am. Spec. Pap.*, **I**, 93–107.

Cameron, E. N. (1969) Post cummulus changes in the Eastern Bushveld Complex, *Am. Mineral.*, **54**, 754–780.

Cameron, E. N. (1971) Problems of the Eastern Bushveld Complex, *Fortsch. Mineral.*, **48**, 86–108.

Campbell, I. H. (1977) A study of macro rhythmic layering and cumulate processes in Jimberlaxa intrusion Western Australia, Part I: The upper layered series, *J. Petrol.*, **18**, 183–215.

Carmichael, I. S. E., and W. S. MacKenzie (1963) Feldspar–liquid equilibria in pantellerites: An experimental study, *Am. J. Sci.*, **261**, 382–396.

Carmichael, I. S. E., and J. Nicholls (1967) Iron titanium oxides and oxygen fugacities in volcanic rocks, *J. Geophys. Res.*, **72**, 4665–4687.

Carmichael, I. S. E., F. J. Turner, and J. Verhoogen (1974) *Igneous Petrology*, McGraw-Hill, New York.

Carswell, D. A. and J. B. Dawson (1970) Garnet peridotite xenoliths in South Africa kimberlite pipes and their petrogenesis, *Contrib. Mineral. Petrol.*, **25**, 163–184.

Carter, J. L. (1970) Mineralogy and chemistry of the earth's upper mantle based on the partial fusion-partial crystallization model, *Bull. Geol. Soc. Am.*, **81**, 2021–2034.

Cawthorn, R. G. (1975) Degrees of melting in mantle diapirs and the origin of ultrabasic liquids, *Earth Planet Sci. Lett.*, **27**, 113–120.

Chayes, F. (1960) On correlation of variables of constant sum, *J. Geophys. Res.*, **65**, 4185–4193.

Chayes, F. (1962) Numerical correlation and petrographical variation, *J. Geol.*, **60**, 440–452.

Christensen, N. I. (1966) Elasticity of ultrabasic rocks, *J. Geophys.*, **70**, 6147–6164.

Clark, P. D., and W. J. Wadsworth (1970) The Insche layered intrusion, *Scot. J. Geol.*, **6**, 7–25.

Clark, S. P., and A. E. Ringwood (1964) Density distribution and constitution of the mantle, *Rev. Geophys.*, **2**, 35–88.

Clark, S. P. Jr., J. F. Schairer, and J. de Neufville (1962) Phase relations in the system $CaMgSi_2O_6$–$CaAl_2Si_2O_8$–SiO_2 at low and high pressures, *Carnegie Inst. Washington Yearb.*, **61**, 59–68.

Cooke, D. L., and W. W. Moorehouse (1969) Timiskaming volcanism in Kirkland area, Ontario, Canada, *Can. J. Earth Sci.*, **6**, 117–132.

Dachin, R. V., and F. R. Boyd (1976) Ultramafic nodules from the Premier kimberlite pipe, South Africa, *Carnegie Inst. Washington Yearb.*, **75**, 531–538.

Daly, R. A. (1911) The nature of the volcanic action, *Proc. Am. Acad. Arts Sci.*, **47**, 48–122.

Daly, R. A. (1925) Relation of mountain building to igneous action, *Proc. Am. Philos. Soc.*, **64**, 283–367.

Daly, R. A. (1928) Bushveld igneous complex of the Transvaal, *Bull. Geol. Soc. Am.*, **39**, 703–768.

Davis, B. T. C., and J. F. Schairer (1965) Melting relations in the join diopside–forsterite–pyrope at 40 kilobar and at one atmosphere, *Carnegie Inst. Washington Yearb.*, **64**, 123–126.

Dawson, J. B. (1962) The geology of Oldoinyo Lengai, *Bull. Volcanol.*, **24**, 155–168.

Dawson, J. B. (1968) Recent Researches on kimberlite and diamond geology, *Econ. Geol.*, **63**, 504–511.

Dawson, J. B., and D. G. Powell (1969) Mica in the upper mantle, *Contrib. Mineral. Petrol.*, **22**, 233–237.

De, A., (1974) Silicate liquid immiscibility in the Deccan traps and its petrogenetic significance, *Bull. Geol. Soc. Am.*, **85**, 471–474.

Dickey, J. S., and H. S. Yoder, Jr., and J. F. Schairer (1971) Chromium in silicate–oxide systems, *Carnegie Inst. Washington Yearb.*, **70**, 118–125.

Dorman, J., and M. Ewing (1962) Numerical inversion of seismic surface wave dispersion data and crust mantle structure in the New York–Pennsylvania area, *J. Geophys. Res.*, **76**, 2587–2601.

Doyle, H. A., and I. Evringham (1964) Seismic velocities and crustal structure in southern Australia, *J. Geol. Soc. Aust.*, **11**, 141–150.

Drever, H. I. (1960) Immiscibility in the picritic intrusions at Igdlorssuit, West Greenland, 21st Int. Geol. Congr., Copenhagen, Pt. 13, pp. 47–58.

Eaton, J. P., and K. J. Murata (1960) How volcanoes grow, *Science*, **132**, 925–938.

Edgar, A. D. (1964) Phase equilibrium studies in the system $CaMgSi_2O_6$ (diopside)–$NaAlSiO_4$ (nepheline)–$NaAlSi_3O_8$ (albite)–H_2O at 1000 kb/cm^2 water vapor pressure, *Am. Mineral.*, **49**, 573–585.

Edgar, A. D., and J. Nolan (1966) Phase relation in the system $NaAlSi_3O_8$ (albite)–$NaAlSiO_4$

References

(nepheline)–$NaFeSi_2O_6$ (acmite)–$CaMgSi_2O_6$ (diopside)–H_2O and its importance in the genesis of alkaline undersaturated rocks, *Ind. Mineral.*, **I.M.A. Vol.**, 176–181.

Eggler, D. H. (1973) Role of CO_2 in the melting processes in the mantle, *Carnegie Inst. Washington Yearb.*, **72**, 257–267.

Eggler D. H. (1974) Effect of CO_2 on the melting of peridotite, *Carnegie Inst. Washington Yearb.*, **73**, 215–224.

Eggler, D. H. (1975a) Peridotite–carbonate relations in the system CaO–MgO–SiO_2–CO_2, *Carnegie Inst. Washington Yearb.*, **74**, 468–474.

Eggler, D. H. (1975b) CO_2 as a volatile component of the mantle: The system Mg_2SiO_4–SiO_2–H_2O–CO_2, *Phys. Chem. Earth*, **9**, 869–881.

Eggler D. H. (1977) The principle of the zone of invariant composition: An example in the system CaO–MgO–SiO_2–CO_2–H_2O and implications for the mantle solidus, *Carnegie Inst. Washington Yearb.*, **76**, 428–435.

Eggler, D. H. (1978) The effect of CO_2 upon partial melting of peridotite in the system Na_2O–CaO–Al_2O_3–MgO–SiO_2–CO_2 to 35 kb, with an analysis of melting in a peridotite–H_2O–CO_2 system, *Am. J. Sci.*, **278**, 305–343.

Eggler, D. H., and C. W. Burnham (1973) Crystallization and fractionation trends in the system andesite–H_2O–CO_2–O_2 at pressures to 10 kb, *Bull. Geol. Soc. Am.*, **84**, 2517–2532.

Eggler, D. H., I. Kushiro, and J. R. Holloway (1976) Stability of carbonated minerals in a hydrous mantle, *Carnegie Inst. Washington Yearb.*, **75**, 631–636.

Eggler, D. H., B. O. Mysen and M. G. Seitz (1974) The solubility of CO_2 in silicate liquids and crystals, *Carnegie Inst. Washington Yearb.*, **73**, 226–228.

Egorov, L. S. (1970) Carbonatites and ultrabasic alkaline rocks of the Maimecha–Kotui region, N. Siberia, *Lithos*, **3**, 341–359.

Ehler, E. G. (1972) *Interpretation of Geological Phase Diagrams*, W. H. Freeman, San Francisco, Calif.

El Gorsey, A., and H. S. Yoder, Jr. (1973) Natural and synthetic mililite compositions, *Carnegie Inst. Washington Yearb.*, **72**, 359–371.

Eskola, P. O. (1932) On the origin of granitic magmas, *Tschermaks Mineral. Petrograph. Mitt.*, **42**, 445–481.

Evans, B. W., and T. L. Wright (1972) Composition of liquidus chromite from the 1959 (Kilauea-Iki) and 1965 (Makaopuhi) eruptions of Kilauea volcano, Hawaii, *Am. Mineral.*, **57**, 217–230.

Fenner, C. N. (1948) Immiscibility of igneous magmas, *Am. J. Sci.*, **246**, 816–850.

Ferguson, J. (1964) Geology of the Ilimaussaq alkaline intrusion, South Greenland, *Med. Groenland*, **1972**, Nr. 4, 82 pp.

Ferguson, J., and K. L. Currie (1971) Evidence of liquid immiscibility in alkaline ultrabasic dikes at Callender Bay, Ontario, *J. Petrol.*, **12**, 561–585.

Ferguson, J., and K. L. Currie (1972) Silicate immiscibility in the ancient "basalts" at the Barberton Mountain land, Transval, *Nature*, **235**, 86–89.

Ferguson, J. B., and A. F. Buddington (1920) The system gehlenite–akermanite, *Am. J. Sci.*, **50**, 133.

Ferguson, J. B., and H. E. Merwin (1919) The ternary system CaO–MgO–SiO$_2$, *Am. J. Sci.,* 48–123.

Fischer, R. (1950) Entmischungen in Schmelzen aus Schwermetalloxyden, Silikaten und Phosphaten, Ihre geochemische und lagustrattenkundiche Bedeutung, *Neus-Jahrb. Mineral.,* **81,** 315–364.

Forbes, R. B., and H. Kuno (1965) The regional petrology of peridotite inclusions and basaltic host rocks, *Proc. Int. Union Geol. Sci., Upper Mantle Symp., New Delhi, 1964,* pp. 161–179.

Ford, C. E. (1972) The system NaFeSi$_2$O$_6$–KAlSi$_3$O$_8$–Na$_2$O·4SiO$_2$–H$_2$O at 1 kbar pressure, *Prog. Exp. Petrol. Nat. Environ. Res. Counc.,* Ser. D, 161–164.

Franco, R. R., and J. F. Schairer (1950) Liquidus temperatures in mixtures of the feldspars of soda, potash and lime, *J. Geol.,* **59,** 259–267.

Fudali, R. F. (1963) Experimental studies bearing on the origin of pseudoleucite and associated problems of alkalic rock systems, *Bull. Geol. Soc. Am.,* **74,** 1101–1126.

Fudali, R. F. (1965) Oxygen fugacities of basaltic and andesitic magmas, *Geochim. Cosmochim. Acta,* **29,** 1063–1075.

Geijer, P. (1931) The iron ores of Kiruna type, geographical distribution, geological characters and origin, *Sver. Geol. Unders.,* 367 pp.

Gerassimovsky, V. I. (1956) Geochemistry and mineralogy of nepheline syenite intrusions, *Geokhimiya* (English summary), **5,** 494–510.

Gerassimovsky, V. I. (1965) Role of F and other volatiles in alkaline rocks, *Geochem. Int.,* **2,** 9.

Gilluly, J. (1971) Plate tectonics and magma evolution, *Bull. Geol. Soc. Am.,* **82,** 2383–2396.

Gold, D. P. (1966) The average and typical chemical composition of carbonatites, *Ind. Mineral.,* **I.M.A. Vol.,** 83–91.

Goldschmidt, V. M. (1930) Element und Mineral pegmatitischer Gesteine, *Nachr. Ges. Wiss. Gottingen Math Phys. Kl.,* 378.

Goranson, R. W. (1931) Solubility of water in granitic magmas, *Am. J. Sci.,* **22,** 481–502.

Goranson, R. W. (1937) Silicate water systems: The osmotic pressure of silicate melts, *Am. Mineral.,* **22,** 485–490.

Goranson, R. W. (1938) Silicate–water systems: Phase equilibria in the NaAlSi$_3$O$_8$–H$_2$O and KAlSi$_3$O$_8$–H$_2$O systems at high temperatures and pressures, *Am. J. Sci., Day Vol.,* **35A,** 71–91.

Graham, E. K., and G. R. Barsch (1969) Elastic constants of single crystal forsterite as a function of temperature and pressure, *J. Geophys. Res.,* **74,** 5949–5960.

Green, D. H. (1964) Petrogenesis of high temperature intrusion in Lizard Area, Cornwall, *J. Petrol.,* **5,** 134–188.

Green, D. H. (1970) The origin of basaltic and nephelinitic magmas, *Trans. Leicester Lit. Philos. Soc.,* **64,** 28–54.

Green, D. H. and R. C. Liebermann (1976) Phase equilibria and elastic properties of a pyrolite model for the oceanic upper mantle, *Tectonophysics,* 3.

Green, D. H. (1973a) Experimental melting studies on a model upper mantle composition at high pressure under water-saturated and water-undersaturated conditions, *Earth Planet. Sci. Lett.,* **19,** 37–53.

References

Green D. H. (1973b) Conditions of melting of basanite magma from garnet peridotite, *Earth Planet. Sci. Lett.*, **197**, 456–465.

Green, D. H. (1975) Genesis of Archean peridotitic magmas and constraints on Archean geothermal gradients and tectonics, *Geology*, **3**, 15–18.

Green, D. H. and R. C. Liebermann (1976) Phase equilibria and elastic properties of a pyrolite model for the oceanic upper mantle, *Tectonophysics*, 3.

Green, D. H. and A. E. Ringwood (1963) Mineral assemblages in the model for the upper mantle, *J. Geophys. Res.*, **68**, 937–945.

Green, D. H. and A. E. Ringwood (1964) Fractionation of basaltic magmas at high pressures, *Nature*, **201**, 1276–1277.

Green, D. H. and A. E. Ringwood (1966) Origin of the calc-alkali igneous rock suite, *Earth Planet Sci. Lett.*, **3**, 307–316.

Green, D. H. and A. E. Ringwood (1967a) The genesis of basaltic magma, *Contrib. Mineral. Petrol.*, **15**, 103–190.

Green, D. H. and A. E. Ringwood (1967) An experimental investigation of gabbro-eclogite transformation and its petrological application, *Geochem. Cosmochim. Acta*, **31**, 767–833.

Green, D. H. and A. E. Ringwood (1967c) The stability fields of aluminous pyroxene, peridotite and garnet peridotite and their relavence in upper mantle structure, *Earth Planet. Sci. Lett.*, **3**, 151–160.

Green, D. H. and A. E. Ringwood (1970) Mineralogy of peridotitic compositions under upper mantle conditions, *Phys. Earth Planet. Inter.*, **3**, 337–359.

Greig, J. W. (1927) Immiscibility in silicate melts, *Am. J. Sci.*, **13**, 133–154.

Gupta, A. K. (1972) The system forsterite–diopside–akermanite–leucite and its significance in the origin of potassium rich mafic-ultramafic volcanic rocks, *Am. Mineral.*, **57**, 1242–2160.

Gupta, A. K. and E. G. Lidiak (1973) The system diopside–nepheline–leucite, *Contrib. Mineral. Petrol.*, **41**, 231–239.

Guttenburg, B. (1951) *Internal Constitution of the Earth*, Dover, New York, 439 pp.

Hamilton, D. L. (1969) Solid solution of anorthite in alkali felspar at 700°C and 900°C, *Prog. Exp. Petrol., Natl. Environ. Res. Counc., London*, **1**, 51–52.

Hamilton, D. L. and G. M. Anderson (1967) Effects of water and oxygen pressures on the crystallization of basaltic magmas, in *Basalts*, Vol. 1, H. H. Hess and A. Poldervaart, Eds., pp. 445–482.

Hamilton, D. L., and W. S. MacKenzie (1965) Phase equilibrium studies in the system $NaAlSiO_4$ (nepheline)–$KAlSiO_4$ (kalsilite)–SiO_2 (silica)–H_2O, *Mineral. Mag.*, **34**, 214–231.

Hamilton, D. L., C. W. Burnham, and E. F. Osborn (1964) The solubility of water and effects of oxygen fugacity of water contents on the crystallization in mafic magmas, *J. Petrol.*, **5**, 21–39.

Hamilton, W. (1965) Diabase sheets of the Taylor glacier region Victoria land, Antarctica, *U.S. Geol. Surv. Prof. Pap.*, **456B**, 1–71.

Harris, P. G., and A. K. Middlemost (1970) The evolution of kimberlites, *Lithos*, **3**, 79–90.

Hawley, J. E. (1962) The Sudbury ores, their mineralogy and origin, *Can. Mineral.*, **7**, Pt. I.

Heald, E. F., J. Naughton, and I. L. Barnes (1963) The chemistry of volcanic gases, use of equilibrium calculation in the interpretation of volcanic gas samples, *J. Geophys. Res.*, **68**, 545–557.

Hess, H. H. (1938) A primary peridotite magma, *Am. J. Sci.*, **35**, 321–344.

Hess, H. H. (1960) Stillwater igneous complex, Montana, a quantitative mineralogical study, *Geol. Soc. Am. Mem.*, **80**.

Hess, P. C. (1971) Polymer model of silicate melts, *Geochim. Cosmochim. Acta*, **35**, 289–306.

Higazy, R. A. (1954) Trace elements of volcanic ultrabasic potassic rocks of southwestern Uganda and adjoining part of Belgian Congo, *Bull. Geol. Soc. Am.*, **65**, 39–70.

Hill, R. E. T., and A. L. Boetcher (1970) Water in earth's mantle: Melting curves of basalt–water and basalt–water–carbon dioxide, *Science*, **167**, 980–982.

Hodges, F. N. (1973) Solubility of H_2O in forsterite melt at 20 kb, *Carnegie Inst. Washington Yearb.*, **72**, 495–497.

Hodges, F. N. (1974) The solubility of H_2O in silicate melts, *Carnegie Inst. Washington Yearb.*, **75**, 251–255.

Holgate, N. (1954) The role of liquid immiscibility in igneous petrogenesis, *J. Geol.*, **62**, 439–480.

Holloway, J. R., and C. W. Burnham (1972) Melting relations of basalts with equilibrium water pressure less than total pressure, *J. Petrol.*, **13**, 1–29.

Holloway, J. R., C. W. Burnham, and G. L. Millholm (1968) Generation of $H_2O + CO_2$ mixtures for use in hydrothermal experimentation, *J. Geophys. Res.*, **73**, 6598–6600.

Holloway, J. R., and C. F. Lewis (1974) CO_2 solubility in hydrous albite liquid at 5 kbar., *Trans Am. Geophys. Union*, **55**, 483.

Holloway, J. R., B. O. Mysen, and D. H. Eggler (1976) The solubility of CO_2 in liquids in the join $CaO–MgO–SiO_2–CO_2$, *Carnegie Inst. Washington Yearb.*, **75**, 626–631.

Holmes, A. (1930) *Petrographic Methods and Calculations*, Murby, London, 515 pp.

Holmes, A. (1932) The origin of igneous rocks, *Geol. Mag.*, **69**, 543–558.

Holmes, A. (1942) A suite of volcanic rocks from southwest Uganda containing Kalsilite (a polymorph of $KAlSiO_4$), *Mineral. Mag.*, **26**, 197–216.

Holmes, A. (1950) Petrogenesis of katungite and its associate, *Am. Mineral.*, **35**, 772–792.

Holmes, A. (1952) The potash ankaratrite–melaleucitite lavas of Nabugando and Mbuga craters, southwest Uganda, *Trans. Geol. Soc. Edinburgh*, **15**, 187–213.

Holmes, A., and H. F. Harwood (1932) Petrology of the volcanic fields east and southeast of Ruwenzori, Uganda, *Q. J. Geol. Soc. London*, **88**, 370–442.

Holmes, A., and H. F. Harwood (1937) The volcanic area of Bufumbira, Pt. I, II, The petrology of the volcanic area of the volcanic field of Bufumbira, Southwest Uganda, *Mem. Geol. Surv. Uganda*, No. 3, 1–300.

Howland, A. L., J. W. Peoples, and E. Sampson (1936) The stillwater igneous complex, *Montana Bur. Mines Geol. Misc. Contrib.*, **7**.

Humpheries, D. J. (1972) Melting behavior of Reunion igneous rocks, *Prog. Expl. Petrol., Natl. Environ. Counc. Publ. London*, Sr. D, 113–115.

Humpheries, D. J., and K. G. Cox (1972) Melting data for volcanic rocks from Aden, South Arabia, *Prog. Expl. Petrol. Natl. Environ. Res. Counc. Publ. London*, Ser. D. 116.

References

Hussak, E. (1900) Ueber ein leukokrates gemischtes ganggestein aus dem nephelin syenit-gebiete der serrade Cladas, *Brasilian News, Jahrb. Min.*, **1**, 22–28.

Hutchison, C. S. (1974) *Laboratory Methods of Petrographic Techniques*, Wiley, New York.

Hutchison, P., D. K. Paul, and P. G. Harris (1970) Chemical composition of the upper mantle, *Mineral. Mag.*, **37**, 726–729.

Hyndman, D. W. (1969) The development of granitic plutons through anataxis in the northern cordillera, British Columbia, *Geol. Soc. Am. Spec. Pap.*, **212**, 146 pp.

Hyndman, D. W. (1972) *Petrology of Igneous and Metamorphic Rocks*, McGraw-Hill, New York.

Irvine, T. N. (1967) Chromian spinel as petrogenetic indicator, II, Petrologic applications, *Can. J. Earth Sci.*, **4**, 71–103.

Irvine, T. N. (1970) Crystallization sequences in the Muskox intrusion and other layered intrusions. I, Olivine–pyroxene–plagioclase relations, *Geol. Soc. S. Africa, Spec. Publ.*, **1**, 441–426.

Irvine, T. N. (1975) Olivine–pyroxene–plagioclase relations in the system Mg_2SiO_4–$CaAl_2Si_2O_8$–$KAlSi_3O_8$–SiO_2 and their bearing on the differentiation of stratiform intrusions, *Carnegie Inst. Washington Yearb.*, **74**, 492–500.

Irvine, T. N. (1976) Metastable liquid immiscibility and MgO–FeO–SiO_2 fractionation patterns in that system Mg_2SiO_4–Fe_2SiO_4–$CaAl_2Si_2O_8$–$KAlSi_3O_8$–SiO_2, *Carnegie Inst. Washington Yearb.*, **75**, 597–611.

Irvine, T. N. (1977) Chromite crystallization in the join Mg_2SiO_4–$CaMgSi_2O_6$–$CaAl_2Si_2O_8$–$MgCr_2O_4$–SiO_2, *Carnegie Inst. Washington Yearb.*, **76**, 465–472.

Irvine, T. N., and C. H. Smith (1967) Primary oxide minerals in the layered series of Muskox intrusion, in *Ultramafic and Related Rocks*, P. J. Wyllie, Ed., Wiley, New York, pp. 38–49.

Ito, K., and G. C. Kennedy (1967) Melting and phase relations in natural peridotite to 40 kbar, *Am. J. Sci.*, **265**, 519–538.

Jackson, E. D. (1963) Stratigraphic and lateral variation of chromite compositions in Stillwater complex, *Mineral. Soc. Am. Spec. Pap.*, **I**, 46.

Jackson, E. D. (1969) Chemical variation in co-existing chromite and olivine in chromite zones of Stillwater complex, in *Magmatic Ore Deposits*, H. D. B. Wilson, Ed., pp. 41–71.

James, R., and D. L. Hamilton (1969) Phase relations in the system $NaAlSi_3O_8$–$KAlSi_3O_8$–$CaAl_2Si_2O_8$–SiO_2 at 1 kilobar water vapor pressure, *Contrib. Mineral. Petrol.*, **21**, 111–141.

Johannsen, A. (1939) *A Descriptive Petrography of Igneous Rocks*, University of Chicago Press, Chicago, Ill., 318 pp.

Kadik, A. A., and D. H. Eggler (1975) Melt-vapor relations on the join $NaAlSi_3O_8$–H_2O–CO_2, *Carnegie Inst. Washington Yearb.*, **74**, 479–484.

Keith, M. L. (1954) Phase equilibria in the system MgO–Cr_2O_3–SiO_2, *J. Am. Ceramic Soc.*, **37**, 490–496.

Khitarov, N. I., and A. A. Kadik (1973) Water and carbon dioxide in magmatic melts and peculiarities of melting process, *Contrib. Mineral. Petrol.*, **41**, 205–215.

King, B. C. (1949) The Napak area of Karmajoa, Uganda, *Mem. Geol. Surv. Uganda*, **5**, 57 pp.

King, B. C. (1965) Petrogenesis of alkaline rock suites of the volcanic and intrusive centers of eastern Uganda, *J. Petrol.*, **6**, 67–100.

Knight, C. W. (1906) A new occurrence of pseudo-leucite, *Am. J. Sci.*, **21**, 286–295.

Knorr, H. (1932) Differentiation und eruptionfolge in Boehmischen Mittelgebirge, *Mitt. Inst. Min. Petrol.*, **42**, 318–370.

Kogarko, L. N. (1964) Geochemistry of fluorine in alkaline rocks, exemplified by the Lovozero alkaline massif, *Geokhimiya* (English summary), **13**, 119–127.

Kogarko, L. N. (1974) Role of volatiles, *The Alkaline Rocks*, H. Sorensen, Ed., Wiley, New York, pp. 447–484.

Kogarko, L. N., and L. A. Gulyaleva (1965) Geochemistry of halogens in alkalic rocks of Lovozero massif (Kola Peninsula), *Geochem. Int.*, **2**, 729–740.

Kogarko, L. N., and I. D. Rhyabchikov (1961) Dependence of contents of halogens in alkalic compounds in gaseous phase on the chemistry of magma, *Geochem. Int.*, **12**, 1195–1201.

Kostervangroos, A. F., and P. J. Wyllie (1966) Liquid immiscibility in the system $Na_2O-Al_2O_3-SiO_2-CO_2$ at pressures to 1 kilobar, *Am. J. Sci.*, **264**, 234–255.

Kostervangroos, A. F., and P. J. Wyllie (1968) Liquid immiscibility in the join $NaAlSi_3O_8-Na_2CO_3-H_2O$ and its bearing on the origin of carbonatites, *Am. J. Sci.*, **266**, 932–967.

Kostervangroos, A. F., and P. J. Wyllie (1973) Liquid immiscibility in the join $NaAlSi_3O_8-CaAl_2Si_2O_8-Na_2CO_3-H_2O$, *Am. J. Sci.*, **273**, 465–487.

Kullerud, G. (1967) Sulphide studies in P. H. Ableson, Ed., *Researches in Geochemistry*, Wiley, New York, pp. 286–321.

Kuno, H. (1959) Origin of Cenozoic petrographic provinces of Japan and surrounding areas, *Bull. Volcanol.*, **20**, 37–76.

Kuno, H. (1960) High alumina basalts, *J. Petrol.*, **1**, 121–145.

Kuno, H. (1964) Aluminum augite and bronzite in alkali olivine basalt from Taka-Sima north Kyusu, Japan, in *Proc. Symp. Adv. Front. Geol.*, Osmania Univ., India, pp. 205–220.

Kuno, H., and A. I. Aoki (1970) Chemistry of ultramafic nodules and their bearing on the origin of basaltic magmas, *Phys. Earth Planet. Int.*, **3**, 273–301.

Kushiro, I. (1968) Compositions of the magmas formed by partial zone melting of earth's upper mantle, *J. Geophys. Res.*, **73**, 619–634.

Kushiro, I. (1969a) Systems bearing on the melting of the upper mantle under hydrous conditions, *Carnegie Inst. Washington Yearb.*, **68**, 240–245.

Kushiro, I. (1969b) The system forsterite-diopside-silica with or without water at high pressures, *Am. J. Sci., Schairer Vol.*, **267A**, 269–294.

Kushiro, I. (1973) Partial melting of garnet lherzolites from kimberlites at high pressure, in *Lesotho Kimberlites*, P. H. Nixon, Ed., Lesotho National Development Corp., Masern, Lesotho, pp. 294–299.

Kushiro, I. (1974) The system forsterite–anorthite–albite–silica–H_2O at 15 kbar and the genesis of andesite magmas in the upper mantle, *Carnegie Inst. Washington Yearb.*, **73**, 244–248.

Kushiro, I., and A. I. Aoki (1968) Origin of some eclogite inclusions in kimberlites, *Am. Mineral.*, **53**, 1347–1367.

References

Kushiro, I., and J. F. Schairer (1963) New data on the system diopside–forsterite–silica, *Carnegie Inst. Washington Yearb.*, **62**, 95–103.

Kushiro, I., and H. S. Yoder, Jr. (1966) Anorthite–forsterite and anorthite–enstatite reactions and their bearing on the basalt–eclogite transformation, *J. Petrol.*, **7**, 337–362.

Kushiro, I., and H. S. Yoder, Jr. (1974) Formation of ecologite from garnet lherzolite liquidus relations in portion of the system $MgSiO_3$–$CaSiO_3$–Al_2O_3 at high pressures, *Carnegie Inst. Washington Yearb.*, **73**, 256–265.

Lambert, I. B., and P. J. Wyllie (1968) Stability of hornblende and a model for the low velocity zone, *Nature*, **219**, 1240–1241.

Lambert, I. B., and P. J. Wyllie (1970) Low velocity zone in the upper mantle: incipient melting caused by water, *Science*, **169**, 764–766.

Levin, E. M., C. R. Robbins, and H. F. McMurdie (1964) *Phase Diagrams for Ceramists*, The American Ceramic Society, Inc., Columbus, Ohio.

Lindsley, D. H. (1968) Melting of plagioclases at high pressure, *New York St. Museum Sci. Serv. Mem.*, **18**, 39–46.

Loewinson-Lessing, F. J. (1935) On a peculiar type of differentiation represented by variolites of Yalguba, Karelia, *Trans. Inst. Petrol. Ac. Sci. USSR*, **5**, 21–27.

Lombaard, B. V. (1935) On the differentiation and relationships of the rocks of the Bushveld complex, *Trans. Geol. Soc. S. Afr.*, **37**, 5–52.

Luth, W. C. (1967) Studies in the system $KAlSiO_4$–Mg_2SiO_4–SiO_2–H_2O-I, Inferred phase relations and petrologic applications, *J. Petrol.*, **8**, 372–416.

Luth, W. C. (1969) The system $NaAlSi_3O_8$–SiO_2 and $KAlSi_3O_8$–SiO_2 to 20 kb and the relationship between H_2O content, P_{H_2O}, P_{Total} in granitic magmas, *Am. J. Sci., Schairer Vol.*, **267A**, 325–341.

Luth, W. C., R. H. Jahns, and O. F. Tuttle (1964) The granite system at pressures of 4 and 10 kilobars, *J. Geophys. Res.*, **69**, 759–773.

MacDonald, G. A. (1953) Pahoehoe, aa, and blocklava, *Am. J. Sci.*, **251**, 169–191.

MacDonald, G. A. (1972) *Volcanoes*, Prentice-Hall, Englewood Cliffs, N.J., 510 pp.

MacKenzie, W. S. (1972) The origin of trachytes and syenites, *Prog. Exp. Petrol., Natl. Environ. Res. Counc., London*, **2**, 45–50.

MacKenzie, W. S., and S. Rahman (1969) Resorption of plagioclase in the ternary feldspar system. *Prog. Exp. Petrol., Natl. Environ. Res. Council, London*, **1**, 58–63.

MacLean, W. H. (1969) Liquidus phase relations in the FeS–FeO–Fe_3O_4–SiO_2 system and their application to geology, *Econ. Geol.*, **64**, 865–884.

Mao, H. K., and P. M. Bell (1976) High pressure physics; the 1-megabar mark on the ruby R, static pressure scale, *Science*, **191**, 851–852.

Marmo, V. (1967) On the granite problem, *Earth Sci. Rev.*, **3**, 7–29.

Marsh, B. D. (1979) Island arc volcanism, *Am. Sci.*, **67**, 161–172.

Marshall, P. (1914) The sequences of lavas at Northhead Otago, *Q. J. Geol. Soc. London*, **70**, 382–406.

McBirney, A. R. (1969) Compositional variations in the Cenozoic calc-alkaline suites of Central America, in *Proc. Andesite Conf., Bull. Oregon Dept. Geol. Mineral. Ind.*, **65**, 185–189.

McBirney, A. R., and Y. Nakamura (1974) Immiscibility in late stage magmas of the Skaergaard intrusion, *Carnegie Inst. Washington Yearb.*, **73**, 348–352.

McDonald, J. A. (1965) Liquid immiscibility as a factor in chromitite seam formation in the Bushveld Complex, *Econ. Geol.*, **60**, 1674–1685.

McGregor, I. D. (1968) Mafic and ultramafic inclusions as indicators of the depth of origin of basaltic magmas, *J. Geophys. Res.*, **73**, 3737–3745.

McGregor, I. D., and J. L. Carter (1970) The chemistry of clinopyroxenes and garnets of eclogite and peridotitic xenoliths from the Roberts Victor mine, South Africa, *Phys. Earth Planet. Inter.*, **3**, 391–397.

Merril, R. B., J. K. Robertson, and P. J. Wyllie (1970) Melting reactions in the system $NaAlSi_3O_8$–$KAlSi_3O_8$–SiO_2–H_2O to 20 kilobars compared with results from feldspar–quartz–H_2O and rock–H_2O systems, *J. Geol.*, **78**, 558–570.

Minakami, T., and S. Sakuma (1953) Report on volcanic activities and volcanological studies concerning them in Japan during 1948–1951, *Bull. Volcanol.*, **14**, 78–130.

Misch, P. (1949) Metasomatic granitization of batholitic dimensions. *Am. J. Sci.*, **247**, 209–245, 372–406.

Moderski, P. J., and A. L. Boetcher (1973) Phase relationship of phlogopite in the system K_2O–MgO–CaO–Al_2O_3–SiO_2–H_2O to 35 kilobars, a better model for micas in the interior of the earth, *Am. J. Sci.*, **273**, 385–414.

Moore, J. G., and L. Calk (1971) Sulphide spherules in vesicles of dredged pillow basalt, *Am. Mineral.*, **56**, 476–488.

Morse, S. A. (1968a) Syenites, *Carnegie Inst. Washington Yearb.*, **67**, 112–120, 120–126.

Morse, S. A. (1968b) Felspars, *Carnegie Inst. Washington Yearb.* **67**, 120–126.

Morse, S. A. (1969) The Kinglapait layered intrusion, *Labrador, Geol. Soc. Am. Mem.*, **111**, 1–204.

Morse, S. A. (1970) Alkali feldspar with water at 5 kb pressure, *J. Petrol.*, **11**, 221–251.

Muan, A. (1958) Phase equilibria at high temperatures in oxide systems involving changes in oxidation states, *Am. J. Sci.*, **256**, 171–207.

Muan, A., and E. F. Osborn (1956) Phase equilibria at liquidus temperatures in system MgO–FeO–Fe_2O_3–SiO_2, *J. Am. Ceramic Soc.*, **39**, 121–140.

Mysen, B. O. (1973) Melting in hydrous mantle: Phase relations of mantle peridotite with controlled water and oxygen fugacites, *Carnegie Inst. Washington Yearb.*, **72**, 467–478.

Mysen, B. O. (1974) Phase relations of garnet websterite + H_2O to 30 kbar pressure, *Carnegie Inst. Washington Yearb.*, **73**, 237–240.

Mysen, B. O. (1975) Melting of hydrous mantle, II: geochemistry of crystals and liquids formed by anataxis of mantle peridotite at high pressures and high temperatures as a function of controlled activities of water, hydrogen and carbon dioxide, *J. Petrol.*, **16**, 549–593.

Mysen, B. O. (1976) The role of volatiles in silicate melts: Solubility of carbon dioxide and water in feldspar, pyroxene and felspathoid melts at 30 kb and 1625°C, *Am. J. Sci.*, **276**, 969–996.

Mysen, B. O., and A. L. Boetcher (1975) Melting in hydrous mantle, I: Phase relations of natural

References

peridotite at high pressures and temperatures with controlled activities of water, carbon dioxide and hydrogen, *J. Petrol.*, **16**, 520–548.

Mysen, B. O., R. J. Arculus, and D. H. Eggler (1975) Solubility of carbon dioxide in melts of andesite, tholeiite and olivine nephelinite compositions to 30 kbar pressure, *Contrib. Mineral. Petrol.*, **53**, 227–239.

Mysen, B. O., M. G. Seitz, and J. D. Franz (1974) Measurements of the solubility of carbon dioxide in silicates utilizing maps of carbon-14 beta activity, *Carnegie Inst. Washington Yearb.*, **73**, 224–226.

Mysen, B. O., D. H. Eggler, M. G. Seitz, and J. R. Holloway (1976) Carbon dioxide in silicate melts and crystals, Part I: Solubility measurements, *Am. J. Sci.*, **276**, 455–479.

Nakamura, Y. (1974) The system Fe_2SiO_4–$KAlSi_2O_6$–SiO_2 at 15 kbar, *Carnegie Inst. Washington Yearb.*, **73**, 352–354.

Naldrett, A. J. (1969) A portion of the system Fe–S–O between 900 and 1080°C and its application to sulphide ore magmas, *J. Petrol.*, **10**, 171–201.

Naslund, H. R. (1976) Liquid immiscibility in the system $KAlSi_3O_8$–$NaAlSi_3O_8$–FeO–Fe_2O_3–SiO_2 and its application to natural magmas, *Carnegie Inst. Washington Yearb.*, **75**, 592–597.

Naslund, H. R. (1977) An investigation of liquid immiscibility in the system K_2O–CaO–FeO–Fe_2O_3–Al_2O_3–SiO_2, *Carnegie Inst. Washington Yearb.*, **76**, 407–410.

Nixon, P. H., O. Vonknorring, and J. M. Rooke (1963) Kimberlites and associated inclusions of Basutoland: A mineralogical and geochemical study, *Am. Mineral.*, **48**, 1090–1132.

Nockolds, S. R. (1954) Average chemical compositions of some igneous rocks, *Bull. Geol. Soc. Am.*, **65**, 1007–1032.

Nolan, J. (1966) Melting relations in the system $NaAlSi_3O_8$–$NaAlSiO_4$–$NaFeSi_2O_6$–$CaMgSi_2O_6$–H_2O and their bearing on the genesis of alkaline undersaturated rocks, *Q. J. Geol. Soc. London*, **122**, 119–157.

Nordlie, B. E. (1971) The composition of the magmatic gas of Kilauea and its behavior in the near surface environment, *Am. J. Sci.*, **271**, 417–463.

O'Hara, M. J. (1965) Primary magmas and the origin of basalts, *Scot. J. Geol.*, **1**, 19–40.

O'Hara, M. J. (1967) Mineral paragenesis in ultrabasic rocks, in *Ultramafic and Related Rocks*, P. J. Wyllie, Ed., Wiley, New York, pp. 393–403.

O'Hara, M. J. (1968) The bearing of phase equilibrium studies in synthetic and natural systems on the origin and evolution of basic and ultrabasic rocks, *Earth Sci. Rev.*, **4**, 69–133.

O'Hara, M. J. (1971) A mechanism for ocean floor spreading, *Phil. Trans. Soc. London*, **268**, 731.

O'Hara, M. J., and H. S. Yoder, Jr. (1963) Partial melting of the mantle, *Carnegie Inst. Washington Yearb.*, **62**, 66–71.

O'Hara M. J., and H. S. Yoder, Jr. (1967) Formation and fractionation of basic magmas at high pressures, *Scot. J. Geol.*, **3**, 67–117.

Osborn, E. F. (1942) The system $CaSiO_3$–diopside–anorthite, *Am. J. Sci.* **240**, 751–788.

Osborn, E. F. (1957) Role of oxygen pressure in the crystallization and differentiation of basaltic magma, *Am. J. Sci.*, **257**, 619–647.

Osborn, E. F. (1959) Role of oxygen pressure in the crystallization and differentiation of basaltic magma, *Am. J. Sci.*, **259**, 609–647.

Osborn, E. F. (1962) Reaction series for subalkaline igneous rocks based on different oxygen pressure conditions, *Am. Mineral.*, **47**, 211–226.

Osborn, E. F. (1963) Some experimental investigations bearing on the origin of igneous magmas of the earth's crust, *Estud. Geol.*, **19**, 1–7.

Osborn, E. F. (1976) Origin of calc-alkali magma series of Santorini Volcano type in the light of recent phase equilibrium studies, *Proc. Int. Cong., Geotherm. Energy Volcanol. Mediterranean Area, Athens, Greece*, pp. 154–162.

Osborn, E. F. (1978) Change in phase relations in response to change in pressure from 1 atm. to 10 kbar for the system Mg_2SiO_4–Iron oxide–$CaAl_2Si_2O_8$–SiO_2, *Carnegie Inst. Washington Yearb.*, **77**, 784–790.

Osborn, E. F., and R. J. Arculus (1975) Phase relations in the system Mg_2SiO_4–iron oxide–$CaAl_2Si_2O_8$–SiO_2 at 10kb and their bearing on the origin of andesite, *Carnegie Inst. Washington Yearb.*, **74**, 504–507.

Osborn, E. F., and A. Muan (1960) Phase diagrams of oxide systems, the system CaO–MgO–SiO_2 and CaO–FeO–SiO_2, American Ceramic Society, Columbus, Ohio, Plates 2 and 7.

Osborn, E. F., and J. F. Schairer (1941) The ternary system pseudowollastonite–akermanite–gehlenite, *Am. J. Sci.*, **239**, 715–763.

Osborn, E. F., and T. Tait (1962) The system diopside-forsterite–anorthite, *Am. J. Sci., Bowen Vol.*, **251A**, 413–433.

Osborn, E. F., and E. B. Watson (1977) Studies of phase relations in subalkaline volcanic rock series, *Carnegie Inst. Washington Yearb.*, **76**, 472–478.

Oxburgh, E. R. (1964) Petrological evidence for presence of amphibole in the upper mantle and its petrological significance and geophysical implications, *Geol. Mag.*, **101**, 1–19.

Patterson, M. S. (1958) Melting of calcite in the presence of water and carbon dioxide, *Am. Mineral.*, **43**, 603–606.

Pearce, M. L. (1964) Solubility of carbon dioxide and variation of oxygen ion activity in soda silica melts, *J. Am. Ceram. Soc.*, **47**, 342–347.

Peck, D. L., J. G. Moore, and G. Kojima (1964) Temperatures in the crust and melt of the Alae Lava Lake Hawaii, after the 1963 eruption of Kilauea volcano, a preliminary report, *U.S. Geol. Surv. Prof. Pap.*, 501-D.

Pecora, W. T. (1956) Carbonatites, A review, *Bull. Geol. Soc. Am.*, **67**, 1537–1556.

Perret, F. A. (1924) The Vesuvius eruption of 1906, *Carnegie Inst. Washington Publ. 339*, 151 pp.; also *Publ. 549*, 162 pp.

Phillips, B., and A. Muan (1959) Phase equilibria in the system CaO–iron oxide–SiO_2 in air, *J. Am. Ceram. Soc.*, **42**, 413–423.

Philpotts, A. R. (1968) Igneous structures and mechanism of emplacement of Mount Johnson, a Monteregion intrusion, Quebec, *Can. J. Earth Sci.*, **5**, 1131–1137; **7**, 195–197.

Philpotts, A. R. (1970) Mechanism of emplacement of Monteregion intrusion, *Can. Mineral.*, **10**, 395–410.

Philpotts, A. R. (1971) Immiscibility between feldspathic and gabbroic magmas, *Nature*, **229**, 107–109.

References

Philpotts, A. R. (1976) Silicate liquid immiscibility: Its probable extent and petrogenetic significance, *Am. J. Sci.*, **276,** 1147–1177.

Philpotts, A. R., and C. J. Hodgson (1968) Role of liquid immiscibility in alkaline rock genesis, *Proc. 23rd Int. Geol. Cong., Paragua, Rept. Sess.*, **2,** 175–188.

Piotrowski, J. M., and A. D. Edgar (1970) Melting relations of alkaline rocks from South Greenland compared to those of Africa and Canada, *Medd. Groenland*, **181,** No. 11, 62 pp.

Platt, R. G., and A. D. Edgar (1969) Phase relations in the join diopside–nepheline–sanidine, *Trans. Am. Geophys. Union*, 50, 337.

Platt, R. G., and A. D. Edgar (1972) The system nepheline–diopside–sanidine and its significance in the genesis of melilite and olivine-bearing alkaline rocks, *J. Geol.*, **80,** 224–236.

Polanski, A. (1949) The alkaline rocks of eastern N. European Plateau, *Bull. Soc. Amis Sci. Let. Poznan, Ser.* B, **10,** 119–184.

Popkov, V. F. (1946) On the activity of Biliukai in 1938–1939, *Bull. Kamchatka Volcano Stn.*, **12,** 29–33.

Presnall, D. C. (1966) The join diopside–forsterite–iron oxide and its bearing on the crystallization of basaltic and ultramafic magmas, *Am. J. Sci.*, **264,** 753–809.

Presnall, D. C. and P. C. Bateman (1973) Fusion relations in the system $NaAlSi_3O_8$–$CaAl_2Si_2O_8$–$KAlSi_3O_8$–SiO_2–H_2O and generation of granitic magmas in the Sierra Nevada batholith, *Bull. Geol. Soc. Am.*, **84,** 3181–3204.

Press, F. (1968) Earth models obtained by Monte Carlo inversion, *J. Geophys. Res.*, **73,** 5223–5234.

Press, F. (1970) Earth models consistent with geophysical data, *Phys. Earth Planet Inter.*, **3,** 3–22.

Press, F., and R. Siever (1978) *Earth*, W. H. Freeman, San Francisco, Calif.

Prince, A. T. (1943) The system albite–anorthite–sphene, *J. Geol.*, **51,** 1–16.

Raguin, E. (1965) *Geology of Granite*, Wiley, New York.

Ramberg, H. (1944) Facies classification of rocks, a clue to the origin of quartzofeldspathic massifs and veins, *J. Geol.*, **57,** 18–54.

Ramberg, H. (1972) Magma diapirism and its tectonic and magma genetic consequences, *Phys. Earth Planet Inter.*, **5,** 45–60.

Rankama, I., and Th. G. Sahama (1950) *Geochemistry*, University of Chicago Press, Chicago, Ill.

Read, H. H. (1957) *The Granite Controversy*, Murby, London.

Read, H. H., M. S. Sadashiaviah, and B. J. Haq (1961) Differentiation in olivine gabbro of the Insche Mass, Aberdeenshire, *Proc. Geol. Assoc.*, pp. 391–413.

Ricci, J. E. (1966) *Phase Rule and the Hetrogeneous Phase Equilibrium*, Dover, New York.

Rickwood, P. C., M. Mathias, and J. C. Seibert (1968) A study of garnets from eclogite and peridotite xenoliths found in kimberlite, *Contrib. Mineral. Petrol.*, **19,** 271–301.

Ringwood, A. E. (1958) Constitution of mantle III; Consequences of the olivine–spinel transition, *Geochim. Cosmochim. Acta*, **15,** 195–212.

Ringwood, A. E. (1962) Model for the upper mantle, *J. Geophys, Res.*, **67,** 857–866, 4473–4477.

Ringwood, A. E. (1966) Mineralogy of the mantle, in *Advances in Earth Sciences*, in P. M. Hurley, Ed., M.I.T. Press, Cambridge, Mass., pp. 357–399.

Ringwood, A. E. (1969) Composition of the crust and upper mantle, in *The Earth's Crust and Upper Mantle*, Vol. 13, P. J. Hart, Ed., American Geophysical Union, Monograph, pp. 1–17.

Ringwood, A. E. (1975) *Composition and Petrology of the Earth's Mantle*, McGraw-Hill, New York.

Rittman, A. (1933) Die geologische bedingte Evolution und Differentiation des Somma-Vesu-Magma, *Z. Vulkanal.*, **15**, 8–94.

Rittman, A. (1962) *Volcanoes and Their Activity*, Wiley, New York, 305 pp.

Roedder, E. (1951) Low temperature liquid immiscibility in the system K_2O–FeO–Al_2O_3–SiO_2, *Am. Mineral.*, **36**, 282–286.

Roedder, E. (1956) Role of liquid immiscibility in igneous petrogenesis. A discussion, *J. Geol.*, **64**, 84–88.

Roedder, E. (1959) Silicate melt systems, *Phys. Chem. Earth*, **3**, 224–297.

Roedder, E. (1965) Liquid CO_2 inclusions in olivine bearing nodules and pheocrysts in basalts, *Am. Mineral.*, **50**, 1746–1782.

Roedder, E., and D. S. Coombs (1967) Immiscibility in granitic melts indicated by fluid inclusions in ejected granitic blocks from Ascension Island, *J. Petrol.*, **8**, 417–451.

Roedder, E., and P. W. Wieblen (1970) Silicate liquid immiscibility in lunar magmas, evidenced by melt inclusions in lunar rocks, *Science*, **167**, 641–644.

Roedder, E., and P. W. Wieblen (1971) Petrology of silicate melt inclusions in Apollo 11 and 12 and terrestial equivalents, *Proc. 2nd Lunar Sci. Conf., Geochim. Cosmochim. Acta Suppl.*, 2, Vol. 1, M.I.T. Press, Cambridge, Mass., pp. 507–528.

Roedder, E., and P. W. Wieblen (1972) Petrographic features and petrologic significance in melt inclusions in Apollo 14 and Apollo 15 rocks, *Proc. 3rd Lunar Sci. Conf., Geochim. Cosmochim. Acta, Suppl. 3*, Vol. 1, M.I.T. Press, Cambridge, Mass., pp. 251–259.

Roeder, P. L. (1960) Phase relations in the Mg_2SiO_4–$CaAl_2Si_2O_8$–FeO–Fe_2O_3–SiO_2 system and their bearing on crystallization of basaltic magma, Ph.D. Thesis, Pennsylvania State University, 104.

Roeder, P. L. (1974) Paths of crystallization and fusion in systems showing ternary solid solution, *Am. J. Sci.*, **274**, 48–60.

Roeder, P. L., and E. F. Osborn (1966) Experimental data for the system MgO–FeO–Fe_2O_3–$CaAl_2Si_2O_8$–SiO_2 and their petrologic implications, *Am. J. Sci.*, **264**, 428–480.

Rosenhauer, M., and D. H. Eggler (1975) Solubility of H_2O and CO_2 in diopside melt, *Carnegie Inst. Washington Yearb.*, **74**, 474–479.

Ross, C. J., M. D. Foster, and A. T. Myers (1954) Origin of dunites and olivine rich inclusions in basaltic rocks, *Am. Mineral.*, **39**, 693–737.

Rubey, W. W. (1951) Geological history of sea water, *Bull. Geol. Soc. Am.*, **62**, 1111–1147.

Sacks, I. S., and H. Okada (1974) A comparison of the anelasticity structure beneath western South America and Japan, *Phys. Earth Planet. Inter.*, **9**, 211–219.

Saggerson, E. P., and L. A. J. Williams (1964) Ngurumanite from southern Kenya and its bearing on the origin of rocks of northern Tanganyika alkaline district, *J. Petrol.*, **5**, 40–81.

Sahama, Th. G. (1960) Kalsilite in lavas of Mt. Nyiragongo (Belgian Congo), *J. Petrol.*, **1**, 146–171.

References

Sahama, Th. G. (1962) Petrology of Mt. Nyiragongo, *Trans. Edinburgh Geol. Soc.*, **19**, 1–28.

Sahama, Th. G. (1974) Potassium-rich alkaline rocks, in *The Alkaline Rocks*, H. Sorensen, Ed., Wiley, New York.

Sahama, Th. G. (1976) Composition of clinopyroxene and melilite in the Nyiragongo rocks, *Carnegie Inst. Washington Yearb.*, **75**, 585–592.

Sahama, Th. G., and A. Myer (1958) A study of the volcano Nyiragongo, A progress report, Institute de Parc Nationaux du Congo Belge, Exploration du Parc National Albert, Missions de Etudes Volcanologiques, Fasc. 2, 1–85.

Scarfe, C. M., W. C. Luth, and O. F. Tuttle (1966) An experimental study bearing on the absence of leucite in plutonic rocks, *Am. Mineral.*, **51**, 726–736.

Schairer, J. F. (1942) The system CaO-FeO-Al_2O_3 I. Results of quenching experiments on five joins, *J. Am. Ceram. Soc.*, **25**, 241–274.

Schairer, J. F. (1950) The alkali feldspar join in the system $NaAlSiO_4$–$KAlSiO_4$–SiO_2, *J. Geol.*, **58**, 512–517.

Schairer, J. F. (1954) The system K_2O–MgO–Al_2O_3–SiO_2 I. Results of quenching experiments on four joins in the tetrahedron cordierite–forsterite–leucite–silica and on the join cordierite–mullite–potash feldspar, *J. Am. Ceram. Soc.*, **37**, 501–533.

Schairer, J. F. (1957) Melting relations of common rock forming minerals, *J. Am. Ceram. Soc.*, **40**, 215–235.

Schairer, J. F. (1959) Phase equilibria with particular reference to silicate systems, in *Physicochemical Measurements at High Temperatures*, J. O'M. Bockris, J. L. White, and J. D. MacKenzie, Eds., Butterworth, London.

Schairer, J. F. (1967) Phase equilibria at one atmosphere related to the tholeiitic and alkali basalts, in *Researches in Geochemistry*, P. H. Abelson, Ed., Wiley, New York, 663 pp.

Schairer, J. F., and N. L. Bowen (1935) Preliminary report on equilibrium relations between feldspathoids, alkali-feldspars and silica, *Trans. Am. Geophys. Union*, 325–328.

Schairer, J. F., and N. L. Bowen (1938) The system diopside–leucite–silica, *Am. J. Sci.*, **35-A**, 289–309.

Schairer, J. F., and N. L. Bowen (1942) The binary system wollastonite–diopside and the relations between $CaSiO_3$-akermanite, *Am. J. Sci.*, **240**, 725–742.

Schairer, J. F., and N. L. Bowen (1947) The system anorthite–leucite–silica, *Bull. Soc. Geol. Finland*, **20**, 67–87.

Schairer, J. F., and N. Morimoto (1958) Systems with rock forming olivines, pyroxenes and feldspars, *Carnegie Inst. Washington Yearb.*, **57**, 212–213.

Schairer, J. F., and N. Morimoto (1959) The system forsterite–diopside–silica–albite, *Carnegie Inst. Washington Yearb.*, **58**, 97–98.

Schairer, J. F., and H. S. Yoder, Jr. (1960a) The nature of residual liquids from crystallization, with data on the system nepheline–diopside–silica, *Am. J. Sci., Bradley Vol.*, **258A**, 273–283.

Schairer, J. F., and H. S. Yoder, Jr. (1960b) The system forsterite–nepheline–diopside, *Carnegie Inst. Washington Yearb.*, **59**, 70–71.

Schairer, J. F., and H. S. Yoder, Jr. (1961) Crystallization in the system nepheline–forsterite–silica at one atmosphere pressure, *Carnegie Inst. Washington Yearb.*, **60**, 141–144.

Schairer, J. F., and H. S. Yoder, Jr. (1964) Crystal and liquid trends in the simplified alkali basalts, *Carnegie Inst. Washington Yearb.*, **63**, 65–74.

Schairer, J. F., and H. S. Yoder, Jr. (1967) The system albite-anorthite-forsterite at 1 atmosphere, *Carnegie Inst. Washington Yearb.*, **66**, 204–209.

Scheumann, K. H. (1913) Petrographische unterSchugen an Gesteinen des Polzen Gebietes im Nord Boehmen, *Mitt. Inst. Mineral. Petrol. Univ. Leipzig*, **34**, 607–776.

Scholtz, D. L. (1936) The magmatic nickeliferous ore deposits of East Grigualand and Pondoland, *Trans. Proc. Geol. Soc. S. Afr.*, **39**, 81–210.

Sederholm, J. J. (1926) On migmatites and associated Pre-cambrian rocks of southwestern Finland, Part II. The region around the Barsundsfjord west of Helsinfors and neighboring areas, *Bull. Comm. Geol., Finl.*, **77**, 143 pp.

Seki, Y., and G. C. Kennedy (1964) The breakdown of potassium feldspar $KAlSi_3O_8$, at high temperatures and high pressures, *Am. Mineral.*, **49**, 1688–1704.

Shand, S. J. (1943) *Eruptive Rocks*, 2nd ed., Wiley, New York.

Shaw, H. R. (1964) Theoretical solubility of water in silicate melts-quasicrystalline model, *J. Geol.*, **72**, 601–617.

Shepard, I. S., G. A. Rankin, and F. E. Wright (1909) The system of alumina with silica, lime and magnesia, *Am. J. Sci.*, **28A**, 293–333.

Sigurdsson, H., and J. G. Schilling (1976) Spinels in mid-Atlantic ridge basalts: Chemistry and occurrence, *Earth Planet. Sci. Lett.*, **29**, 7–20.

Sigvaldsson, G. E. and G. Elisson (1968) Collection and analysis of volcanic gases at Surtsey, Iceland, *Geochim. Cosmochim. Acta*, **32**, 797–805.

Smith, F. G. (1961) Metallic sulphide melts as igneous differentiates, *Can. Mineral.*, **6**, 663–669.

Smith, W. C. (1956) A review of some problems of African carbonatites, *Q. J. Geol. Soc. London*, **112**, 189–219.

Smyth, C. H. (1927) Origin of alkaline rocks, *Proc. Am. Phil. Soc.*, **66**, 535–580.

Solomon, S. C. (1972) Seismic wave attenuations and partial melting in the upper mantle of North America, *J. Geophys. Res.*, **77**, 1483–1502.

Sood, M. K., and A. D. Edgar (1967) Melting relations in undersaturated alkaline rocks, II, *Can. Mineral.*, **9**, Pt. 2, 308.

Sood, M. K., and A. D. Edgar (1969) Phase relations in the system diopside–albite–leucite, *Trans. Am. Geophys. Union*, **50**, 337.

Sood, M. K., and A. D. Edgar (1970) Melting relations of undersaturated alkaline rocks from the Ilimaussaq Intrusion and Grondal-Ika complex South Greenland under water vapor and controlled partial oxygen pressure, *Medd. Groenland*, **181**, No. 12, 41 pp.

Sood, M. K., and A. D. Edgar (1972) The system diopside–forsterite–nepheline–albite–leucite and its application to the genesis of alkaline rocks, *Proc. 24th Int. Geol. Cong., Montreal, Canada, Sec. 14*, 68–74.

Sood, M. K., and R. Ellis (1973) Electron microscope studies of selected glassy rocks, *Geol. Soc. Am. Abstr.*, **5**, 816.

References

Sood, M. K., P. E. Myers, and L. A. Berlin (1980) Petrology, geochemistry and contract relations of the Wausau and Stettin Plutons, Central Wisconsin, Field Guide, Trip 3, 26th Annual Institute of Lake Superior Geology, 59 pp.

Sood, M. K., R. G. Platt, and A. D. Edgar (1970) Phase equilibrium relations in portions of the system diopside–nepheline–kalsilite–silica and their importance in the genesis of alkaline rocks, *Can. Mineral.*, **10**, 380–394.

Sorensen, H. (1958) The Illimaussaq batholith, A review and discussion, *Medd. Groenland*, **162**, No. 3, 48 pp.

Sorensen, H. (1960) On the agpaitic rocks, *Proc. 21st Int. Geol. Cong. Copenhagen*, Pt. 13, 319–327.

Sorensen, H. (1962) On the occurrence of the Steenstrupine in the Illimaussaq batholith Southwest Greenland, *Medd. Groenland*, **167**, No. 1, 251 pp.

Sorensen, H. (1969) Rhythmic igneous layering in peralkaline intrusions, *Lithos*, **2**, 261–283.

Sorensen, H. (1974) *The alkaline rocks*, Wiley, New York, 622 pp.

Speidel, D. H., and E. F. Osborn (1969) Element distribution among co-existing phases in the system $MgO-FeO-Fe_2O_3-SiO_2$ as a function of temperature and oxygen fugacity, *Am. Mineral.*, **52**, 1139–1152.

Stewart, D. B. (1957) The system $CaAl_2Si_2O_8-SiO_2-H_2O$, *Carnegie Inst. Washington Yearb.*, **56**, 214–216.

Stewart, D. B., and E. H. Roseboom, Jr. (1962) Low temperature terminations of three phase region plagioclase-alkali feldspar-liquid, *J. Petrol.*, **3**, 280–315.

Thompson, R. N., and W. S. MacKenzie (1967) Feldspar liquid equilibria in paralkaline acid liquids: An experimental study, *Am. J. Sci.*, **265**, 714–734.

Thompson, R. N., and C. E. Tilley (1969) Melting and crystallization relations of some Kileauean basalts: The lavas of the 1959–60 Kilauea eruption, *Prog. Exp. Petrol., Natl. Environ. Res. Counc.*, **1**, 32–40.

Tilley, C. E. (1950) Some aspects of magmatic evolution, *Q. J. Geol. Soc. London*, **106**, 37–61.

Tilley, C. E. (1957) The problems of alkali rock genesis, *Q. J. Geol. Soc. London*, **113**, 323–359.

Tilley, C. E., and H. S. Yoder, Jr. (1964) Pyroxene fractionation in mafic magma at high pressures and its bearing on basalt genesis, *Carnegie Inst. Washington Yearb.*, **63**, 114–121.

Tilley, C. E., H. S. Yoder, Jr., and J. F. Schairer (1964) New relations on melting of basalts, *Carnegie Inst. Washington Yearb.* **63**, 92–97.

Tilley, C. E., H. S. Yoder, Jr., and J. F. Schairer (1967) Melting relations of volcanic rock series, *Carnegie Inst. Washington Yearb.*, **66**, 260–269.

Tilley, C. E., H. S. Yoder, Jr., and J. F. Schairer (1968) Melting relations of igneous rock series, *Carnegie Inst. Washington Yearb.*, **67**, 450–457.

Tomkeieff, S. I. (1952) Analcite–trachybasalt inclusions in the phonolite of Traprain Law, *Trans. Geol. Soc. Edinburgh*, **15**, 360–373.

Turner, F. J., and J. Verhoogen (1960) *Igneous and Metamorphic Petrology*, McGraw-Hill, New York.

Tuttle, O. F., and N. L. Bowen (1958) Origin of granite in the light of experimental studies in the system $NaAlSi_3O_8-KAlSi_3O_8-SiO_2-H_2O$, *Geol. Soc. Am. Mem.*, **74**, 153 pp.

Tuttle, O. F., and J. Gittins (1966) *Carbonatites*, Wiley-Interscience, New York.

Tyler, R. C., and B. C. King (1967) The pyroxenes of the alkaline igneous complexes of eastern Uganda, *Mineral. Mag.*, **36**, 5–21.

Ulmer, G. C. (1969) Experimental investigations of chromite spinels, *Econ. Geol., Monogr. No. 4, Magmatic Ore Deposits*, pp. 114–131.

Ussing, N. V. (1912) Geology of the country around Julianehaab, Greenland, *Medd. Groenland*, **38**, 376 pp.

Van Zyl, J. P. (1970) Petrology of the Merensky Reef and associated rocks on Swartklip, 988, Rustenberg District, *Geol. Soc., S. Afr. Spec. Publ.*, **1**, 80–107.

Velde, D., and H. S. Yoder, Jr. (1976) The chemical composition of melilite-bearing eruptive rocks, *Carnegie Inst. Washington Yearb.*, **75**, 574–580.

Velde, D., and H. S. Yoder, Jr. (1977) Melilitite and melilite-bearing igneous rocks, *Carnegie Inst. Washington Yearb.*, **76**, 478–485.

Verhoogen, J. (1939) New data on volcanic gases: The 1938 eruption of Nyamuragira, *Am. J. Sci.*, **237**, 656–672.

Verhoogen, J. (1948) Les eruptions 1938–1940 du Volcan Nyamuragira: Institute de Parc Nationaux du Congo Belge Exploration du Parc National Albert: Mission by J. Verhoogen (1938–1940), Missions d'Etudes Volcanologiques 1, 1948. Fasc. 1, Bruxelles, 186 pp.

Verma, R. K. (1960) Elasticity of some high density crystals, *J. Geophys. Res.*, **65**, 757–766.

Vittaliano, C. J. (1971) Capping volcanic rocks—West Central Nevada, *Bull. Volcanol.*, **34**, 617–635.

Vogt, J. H. L. (1921) The physical chemistry of magmatic differentiation in igneous rocks, *J. Geol.*, **29**, 319–350.

Von Eckerman, H. (1966) Progress of research on Alnö carbonatite, in *Carbonatites*, O. F. Tuttle and J. Gittins, Eds., Wiley, New York, p. 1.

Von Eckerman, H. (1967) A comparison of Swedish, African and Russian kimberlites, in *Ultramafic and Related Rocks*, P. J. Wyllie, Ed., Wiley, New York.

Von Platten, H. (1965) Experimental anataxis and genesis of migmatites, in *Controls of Metamorphism*, W. S. Pitcher and G. W. Flinn, Eds., Oliver and Boyd, Edinburgh, pp. 202–218.

Wager, L. R. (1953) Layered intrusions, Dansk. Geol. Foren. *Medd.*, **12**, 335–349.

Wager, L. R. (1965) The form and internal structure of the alkaline Kangerdlussuaq intrusion, E. Greenland, *Mineral. Mag.*, **34**, 487–493.

Wager, L. R., and G. M. Brown (1957) Funnel shaped intrusions, *Bull. Geol. Soc. Am.*, **68**, 1072.

Wager, L. R., and G. M. Brown (1968) *Layered Igneous Rocks*, Oliver and Boyd, London.

Wager, L. R., and W. A. Deer (1939 reissued in 1962) The petrology of the Skaergaard intrusion, Kangeredlussuaq, E. Greenland, *Medd. Groenland*, **105**, No. 4, 352 pp.

Wager, L. R., E. A. Vincent, and A. A. Smales (1957) Sulphides in the Skaergaard intrusion, E. Greenland, *Econ. Geol.*, **52**, 855–903.

References

Wagner, P. A. (1928) The evidence of the kimberlite pipes on the constitution of the outer part of the earth, *S. Afr. J. Sci.*, **25**, 127–148.

Walker, G. P. L., and R. R. Skelhorn (1966) Some associations of acid and basic igneous rocks, *Earth Sci. Rev.*, **2**, 93–109.

Wang, C. Y. (1970) Density and constitution of the mantle, *J. Geophys. Res.*, **75**, 3264–3284.

Wang, C. Y. (1972) A simple earth model, *J. Geophys. Res.*, **77**, 4318–4329.

Washington, H. S. (1896a) Italian petrological sketches I., The Bolsena region, *J. Geol.*, **4**, 541–566.

Washington, H. S. (1896b) Italian petrological sketches, II., The Viterbo region, *J. Geol.*, **4**, 826–849.

Washington, H. S. (1987) Italian petrological sketches, V. Summary and conclusions, *J. Geol.*, **5**, 349–377.

Watkinson, D. H., and P. J. Wyllie (1964) The limestone assimilation hypothesis, *Nature*, **204**, 1053–1054.

Watson, E. B., and H. R. Naslund (1977) The effect of pressure on liquid immiscibility in the system $K_2O-FeO-Al_2O_3-SiO_2-CO_2$, *Carnegie Inst. Washington Yearb.*, **76**, 410–414.

Wenlandt, R. F. (1977a) The system $KAlSiO_4-Mg_2-SiO_4-SiO_2-CO_2$, Phase relations involving forsterite, enstatite, sanidine, kalsilite, leucite vapor and liquid to 30 kb, *Carnegie Inst. Washington Yearb.*, **76**, 435–441.

Wenlandt, R. F. (1977b) The system $K_2O-MgO-Al_2O_3-SiO_2-H_2O-CO_2$: Stability of phlogopite as a function of vapor composition at high pressures and temperatures, *Carnegie Inst. Washington Yearb.*, **76**, 441–448.

Wilkinson, J. F. G. (1956) Clinopyroxenes of alkali olivine basalt magma, *Am. Mineral.*, **41**, 724–743.

Wilshire, H. G. (1967) The Prospect alkaline diabase-picrite intrusion, New South Wales, Australia, *J. Petrol.*, **8**, 97–163.

Winkler, H. G. F. (1967) *Petrogenesis of Metamorphic Rocks*, 2nd ed., Springer-Verlag, New York.

Worst, B. G. (1958) The differentiation and structure of the great dyke of southern Rhodesia, *Trans. Geol. Soc. S. Afr.*, **59**, 283–358.

Worst, B. G. (1960) The great dyke of southern Rhodesia, *Southern Rhodesia Geol. Surv. Bull. 47*.

Wright, J. B. (1963) A note on possible differentiation trends in Tertiary to recent lavas of Kenya, *Geol. Mag.*, **100**, 164–180.

Wright, J. B. (1971) The phonolite–trachyte spectrum, *Lithos*, **4**, 1–5.

Wyllie, P. J. (1960) The system $CaO-CO_2-H_2O$ and the origin of carbonatites, *J. Petrol.*, **1**, 1–46.

Wyllie, P. J. (1965) Melting relationships in the system $CaO-MgO-CO_2-H_2O$ with petrological applications, *J. Petrol.*, **6**, 101–123.

Wyllie, P. J. (1966) Experimental data on the petrogenetic links between kimberlites and carbonatites, *Ind. Mineral., I.M.A. Vol.*, 67–82.

Wyllie, P. J. (1967) *Ultramafic and Related Rocks*, Wiley, New York.

Wyllie, P. J. (1970) Ultramafic rocks and the upper mantle, *Min. Soc. Am. Spec. Pap.*, **3**, 3–32.

Wyllie, P. J. (1971a) *The Dynamic Earth*, Wiley, New York.

Wyllie, P. J. (1971b) Role of water in magma generation and initiation of diapiric rise in the mantle, *J. Geophys. Res.*, **76**, 1328–1338.

Wyllie, P. J. (1977a) Peridotite–H_2O–CO_2 and carbonatitic liquids in the upper aesthenosphere, *Nature*, **266**, 45–47.

Wyllie, P. J. (1977b) Mantle fluid compositions buffered by carbonates in peridotite–CO_2–H_2O, *J. Geol.*, **85**, 187–207.

Wyllie, P. J., and A. L. Boetcher (1969) Liquidus phase relationships in the system CaO–CO_2–H_2O to 40 kilobars pressure with petrological applications, *Am. J. Sci.*, **267A**, 489–508.

Wyllie, P. J., and J. L. Haas (1965) The system CaO–SiO_2–CO_2–H_2O, I. Melting relationships with excess vapor at 1 kilobar pressure, *Geochim Cosmochim. Acta*, **29**, 871–892.

Wyllie, P. J., and W. L. Huang (1975) Peridotite, kimberlite and carbonatite explained in the system CaO–MgO–SiO_2–CO_2, *Geology*, **3**, 621–624.

Wyllie, P. J., and W. L. Huang (1975) Influence of mantle CO_2 in the generation of carbonatites and kimberlites, *Nature*, **247**, 297–299.

Wyllie, P. J., and W. L. Huang (1976) Carbonation and melting relations in the system CaO–MgO–SiO_2–CO_2 at mantle pressures with geophysical and petrological applications, *Contrib. Mineral. Petrol.*, **54**, 79–107.

Wyllie, P. J., and O. F. Tuttle (1959a) Synthetic carbonatite magma, *Nature*, **183**, 770.

Wyllie, P. J., and O. F. Tuttle (1959b) Effect of carbon dioxide on the melting of granite and feldspars, *Am. J. Sci.*, **257**, 648–655.

Wyllie, P. J., and O. F. Tuttle (1960a) The system CaO–CO_2–H_2O and the origin of carbonatites, *J. Petrol.*, **1**, 1–46.

Wyllie, P. J., and O. F. Tuttle (1960b) Experimental investigations of silicate systems containing two volatile components, I: Geometrical considerations, *Am. J. Sci.*, **258**, 498–517.

Yagi, K. (1953) Petrochemical studies on the alkalic rocks of the Morotu District, Sakhalin, *Bull. Geol. Soc. Am.*, **64**, 769–810.

Yagi, K. (1962) A reconnaissance of the system acmite–diopside and acmite–nepheline, *Carnegie Inst. Washington Yearb.*, **61** 98–99.

Yoder, H. S., Jr. (1954) Synthetic basalt, *Carnegie Inst. Washington Yearb.*, **53**, 106–107.

Yoder, H. S., Jr. (1958) Effect of water on melting of silicates, *Carnegie Inst. Washington Yearb.*, **57**, 189–191.

Yoder, H. S., Jr. (1965) Diopside–anorthite–water at five and ten kilobars and its bearing on explosive volcanism, *Carnegie Inst. Washington Yearb.*, **64**, 82–89.

Yoder, H. S., Jr. (1968) Experimental studies bearing on the origin of anorthosites, *New York State Mus. Sci. Serv. Mem.*, 13–22.

Yoder, H. S., Jr. (1969) Calc-alkali andesites: Experimental data bearing on the origin of their assumed characteristics, *Proc. Andesite Conf. Oregon Dept. Geol. Mineral. Ind. Bull.*, **65**, 77–89.

Yoder, H. S., Jr. (1973a) Contemporaneous basaltic and rhyolitic magmas, *Am. Mineral.*, **58**, 153–172.

References

Yoder, H. S., Jr. (1973b) Melilite stability and paragenesis, *Fortschr. Mineral.*, **50**, 140–173.

Yoder, H. S., Jr. (1973c) Akermanite–CO_2: Relationship of melilite-bearing rocks to kimberlite, *Carnegie Inst. Washington Yearb.* **72**, 449–457.

Yoder, H. S., Jr. (1974) Garnet peridotite as the parental material for basaltic liquids, *Carnegie Inst. Washington Yearb.*, **73**, 263–266.

Yoder, H. S., Jr. (1975) Heat of melting of simple systems related to basalts of eclogites, *Carnegie Inst. Washington Yearb.*, **74**, 515–519.

Yoder, H. S., Jr. (1976) *Generation of Basalt Magmas,* National Acadamy of Science, Washington, D.C., 265 pp.

Yoder, H. S., Jr. and I. Kushiro (1969) Melting of hydrous phase: Phlogopite, *Am. J. Sci. Schairer Vol.*, **267-A**, 558–582.

Yoder, H. S., Jr. and C. E. Tilley (1962) Origin of basalt magmas: An experimental study of natural and synthetic rock systems, *J.Petrol.*, **3**, 342–532.

Yoder, H. S., Jr., and B. G. J. Upton (1971) Diopside–sanidine–H_2O at 5 and 10 kb., *Carnegie Inst. Washington Yearb.* **70**, 112–118.

Yoder, H. S., Jr., and D. Velde (1976) Importance of alkali content of magma yielding melilite-bearing rocks, *Carnegie Inst. Washington Yearb.*, **75**, 580–585.

Yoder, H. S., Jr., D. B. Stewart, and J. R. Smith (1957) Ternary feldspars, *Carnegie Inst. Washington Yearb.*, **56**, 206–214.

Yoshiki, B., and R. Yoshida (1952) Composition of low alkali glasses, *J. Am. Ceram. Soc.*, **35**, 166–169.

Zavaritskii, A. N., and V. S. Sobolev (1964) *The Physicochemical Principle of Igneous Petrology,* Israel Program *Science Translation and Publication, Tel Aviv*.

Zies, E. G. (1941) Temperatures of volcanoes, fumaroles and hot springs, in *Temperature: Its Measurement and Control in Science and Industry,* American Institute of Physics, New York, pp. 372–380.

Author Index

Akella, J., 171, 205
Andersen, O., 6, 17-18, 20-22, 205, 206
Anderson, D. L., 183, 205
Anderson, G. M., 69, 167, 205, 211
Aoki, A. I., 171, 214
Arculus, R. J., 66, 73, 75-76, 205, 217, 218
Arndt, N. T., 153, 184, 205
Asklund, B., 113, 205

Bailey, D. K., 30, 82-83, 105, 109-112, 124, 144, 168, 183, 205
Barker, D. S., 158, 205
Barth, T. F. W., 115, 205
Barnes, H. L., 207
Barnes, I. L., 211
Barsch, G. R., 172, 210
Bateman, P. C., 130, 219
Bell, K., 107, 206
Bell, P. M., 7, 206
Berlin, L. A., 223
Billibin, Y. U. A., 113, 206
Birch, F., 178, 206
Bockris, J. O. M., 148, 206
Boetcher, A. L., 115, 144-145, 147, 153, 206, 216, 226
Borley, G. D., 107, 206
Bowen, N. L., 6-8, 11-18, 20, 22, 46, 66, 72, 77, 80-82, 87, 90, 99-100, 113, 119-125, 130, 137, 142, 152, 169, 171, 175, 180, 206, 221
Boyd, F. R., 171, 174, 205, 206-208
Brey, G., 142, 145, 196, 207
Brotzu, P. L., 105, 207
Brown, G. M., 26, 30, 73, 207, 224
Buddington, A. F., 12, 209
Bullard, F. M., 164, 207
Burnham, C. W., 141, 142-144, 207, 209

Calk, L., 113, 216

Cameron, E. N., 30, 40, 207
Campbell, I. H., 26, 171, 207
Carmichael, I. S. E., 76, 124, 194, 207
Carswell, D. A., 5, 174, 207
Carter, J. L., 171, 207, 216
Cawthorn, R. G., 184, 185, 207
Chayes, F., 163, 208
Christensen, N., 178, 208
Clark, P. D., 134, 208
Clark, S. P., 9, 184, 208
Cooke, D. L., 85, 208
Coombs, D. S., 113, 220
Cox, K. G., 158, 161, 212
Currie, K. L., 113, 209, 210

Dachin, R. V., 171, 208
Daly, R. A., 27, 164, 170, 208
Davis, B. T. C., 193, 195-196, 208
Dawson, J. B., 5, 115, 168, 171, 174, 207, 208
De, A., 113, 208
Deer, W. A., 26, 224
De Neufville, J., 208
Dickey, J. S., 76, 208
Dorman, J., 172, 208
Doyle, H. A., 172, 208
Drever, H. I., 113, 208

Eaton, J. P., 186, 208
Edgar, A. D., 41, 88, 93, 96, 101-106, 134-136, 158-167, 208, 219, 222
Eggler, D. H., 7, 114, 137-138, 145-147, 188, 196, 209, 217, 220
Egorov, L. S., 144, 209
Ehler, E. G., 13, 38, 41, 209
El Gorsey, A., 65, 209
Elisson, G., 115, 222
Ellis, R., 113, 222
Eskola, P. O., 169, 209

229

Evans, B. W., 40, 209
Evringham, I., 172, 208
Ewing, M., 172, 208

Fenner, C. N., 113, 209
Ferguson, J., 113, 166, 168, 209, 210
Ferguson, J. B., 12, 62, 209
Fischer, R., 113, 210
Forbes, R. B., 171, 210
Ford, C. E., 137, 210
Foster, M. D., 171, 220
Franco, R. R., 125, 127, 210
Franz, J. D., 217
Fudali, R. F., 100, 167, 210

Geijer, P., 113, 210
Gerassimovsky, V. I., 158, 164, 167, 210
Gillberg, M. E., 205
Gilluly, J., 169, 210
Gittins, J., 144, 224
Gold, D. P., 144, 210
Goldschmidt, V., 158, 210
Goranson, R. W., 140-142, 210
Graham, E. K., 172, 210
Green, D. H., 7, 115, 142, 145, 153-157, 169, 171, 174-177, 179, 184-185, 188-190, 196, 204, 210, 211
Greig, J. W., 6, 113, 211
Gulyaleva, L., 167, 214
Gupta, A. K., 101, 103, 107, 211
Guttenburg, B., 164, 211

Haas, J. L., 144, 226
Hall, James G., 6
Hamilton, D. L., 69, 100, 125, 128-131, 142, 153, 167, 211
Hamilton, W., 113, 211
Haq, B. J., 219
Harris, P. G., 171, 211, 213
Harwood, H. F., 99, 105, 212
Hawley, J. E., 113, 211
Heald, E. F., 140, 211
Hess, H. H., 27, 30, 172, 212
Hess, P. C., 113, 212
Higazy, R. A., 107, 212
Hill, R. E. T., 115, 145, 212
Hodges, F. N., 142, 212
Hodgson, C. J., 113, 219
Holgate, N., 113, 212

Holloway, J. R., 140-141, 144-145, 209, 212, 217
Holmes, A., 85, 99, 105, 169, 212
Howland, A. L., 26, 212
Huang, W. L., 144-145, 147, 183, 196, 226
Humpheries, D. J., 158, 161, 212
Hussak, E., 85, 213
Hutchison, C. S., 13, 213
Hutchison, P., 171, 213
Hyndman, D. W., 5, 54, 126, 128, 163, 169-170, 174, 213

Irvine, T. N., 7, 21, 24, 25, 26, 29-30, 40, 76, 114, 213
Ito, K., 196, 213

Jackson, E. D., 40, 213
Jahns, R. H., 142, 207, 215
James, R., 125, 213
Johannsen, A., 84, 213

Kadik, A. A., 142, 145, 213
Keith, M. L., 76, 213
Kennedy, G. C., 100, 196, 213
Khitarov, H. I., 142, 213
King, B. C., 54, 83, 213, 214, 224
Knight, C. W., 100, 214
Knorr, H., 54, 105, 214
Kogarko, L. N., 113-114, 165-167, 214
Kojima, G., 218
Kostervangroos, A. F., 113, 144, 214
Kullerud, G., 113, 214
Kuno, H., 169-171, 187, 210, 214
Kushiro, I., 6-7, 46-47, 50, 134-137, 142, 147, 169, 171-173, 175-176, 188-189, 209, 214, 227

Lambert, I. B., 147, 170, 183, 215
Levin, E. M., 13, 113, 215
Lewis, C. F., 140, 212
Lidiak, E. G., 103, 211
Liebermann, R. C., 177-179, 211
Lindsley, D. H., 138, 215
Loewinson-Lessing, F. Von, 113, 215
Lombaard, B. V., 27, 215
Luth, W. C., 123-125, 137, 142, 169, 215, 221

McBirney, A. R., 113, 136, 216

Author Index

MacDonald, G. A., 1, 3, 115, 140, 164, 215
McDonald, J. A., 113, 216
McGregor, I. D., 171, 216
MacKenzie, W. S., 100, 124, 130, 134, 215, 223
MacLean, W. H., 66, 215
McMurdie, H. F., 215
Mao, H. K., 7, 215
Marmo, V., 169, 215
Marsh, B. D., 186, 215
Marshall, P., 113, 215
Mathias, M., 219
Merril, R. B., 158, 216
Merwin, H. E., 62, 209
Middlemost, A. K., 171, 211
Millholm, G. K., 212
Minakami, T., 164, 216
Misch, P., 169, 216
Moderski, P. J., 147, 216
Moore, J. G., 113, 216, 218
Moorehouse, W. W., 85, 208
Morbidelli, P. L., 207
Morey, G. W., 6
Morimoto, N., 35-37, 46-48, 51, 103, 221
Morse, S. A., 86-87, 90-91, 120-122, 125, 130-134, 216
Muan, A., 20, 66, 216, 218
Murata, K. J., 186, 208
Myer, A., 164, 221
Myers, A. T., 220
Myers, P. E., 223
Mysen, B. O., 7, 140, 145, 153, 209, 216, 217

Nakamura, Y., 7, 113, 114, 216, 217
Naldrett, A. J., 113, 217
Naslund, H. R., 7, 114, 217
Naughton, J., 211
Nicholls, J., 76, 207
Nixon, P. H., 171, 174, 217
Nockolds, S. R., 5, 69, 140, 174, 217
Nolan, J., 83, 134-136, 217
Nordlie, B. E., 1, 217

O'Hara, M. J., 7, 106, 136, 169-171, 187-188, 196, 217
Okada, H., 184, 220

Osborn, E. F., 7, 9, 20, 35, 41, 66-76, 153, 205, 211, 217, 218, 220
Oxburgh, E. R., 115, 218

Patterson, M. S., 144, 218
Paul, D. K., 171, 213
Pearce, M. L., 144, 218
Peck, D. L., 164, 218
Pecora, W. T., 144, 218
Peoples, J. W., 212
Perret, F. A., 164, 218
Phillips, B., 66, 218
Philpotts, A., 113, 218
Piotrowski, J., 41, 158, 165, 219
Platt, R. G., 88-93, 96, 219, 223
Polanski, A., 158, 219
Popkov, V. F., 164, 219
Posnjak, E., 206
Powell, D. G., 115, 208
Powell, J. L., 107, 206
Presnell, D. C., 66, 130, 219
Press, F., 172, 181, 219
Prince, A. T., 10, 219

Raguin, E., 169, 219
Rahman, S., 215
Ramberg, H., 169, 184, 219
Rankama, I., 219
Rankin, G. A., 222
Read, H. H., 134, 169, 219
Rhyabchikov, I. D., 113, 165, 214
Ricci, J. E., 41, 109, 219
Rickwood, P. C., 172, 219
Ringwood, A. E., 7, 153-157, 169-179, 174-178, 181, 183-185, 188, 191, 202-204, 208, 211, 219, 220
Rittman, A., 99, 164, 220
Robbins, C. R., 215
Robertson, J. K., 216
Roedder, E., 13, 66, 113, 115, 140, 144, 220
Roeder, P. L., 66, 70, 71-74, 220
Rooke, J. M., 217
Roseboom, E. H., 128, 223
Rosenhauer, M., 145, 220
Ross, C. J., 171, 220
Rubey, W. W., 175, 220

Sacks, I. S., 184, 220

Sadashiaviah, M. S., 219
Saggerson, E. P., 54, 220
Sahama, Th. G., 105-107, 220
Sakuma, S., 164, 216
Sammis, C., 183, 205
Sampson, E., 212
Scarfe, C. M., 100, 221
Schairer, J. F., 6, 12-13, 20, 28, 34-37, 39-41, 48-54, 58, 66, 73, 78-80, 82-83, 87, 90, 100, 103, 105, 109-113, 123, 125, 148, 163, 168, 175, 189, 193-196, 206, 208, 210, 215, 221
Scheumann, K. H., 54, 105, 222
Schilling, J. G., 40, 222
Scholtz, D. L., 113, 222
Sederholm, J. J., 169, 222
Seibert, J. C., 219
Seitz, M. G., 209, 217
Seki, Y., 100, 222
Shand, S. J., 99, 222
Shaw, H. R., 141, 222
Shepard, I. S., 148, 222
Siever, R., 181, 219
Sigurdsson, H., 40, 222
Sigvaldsson, G., 115, 222
Skelhorn, R. R., 113, 225
Smales, A. A., 224
Smith, C. H., 30, 213
Smith, F. G., 113, 222
Smith, J. R., 227
Smith, W. C., 114, 222
Smyth, C. H., 82, 222
Sobolev, V. S., 13, 41, 227
Solomon, S. C., 184, 222
Sood, M. K., 34, 41, 82-83, 86-87, 90-95, 101-106, 113, 134, 158-167, 222, 223
Sorensen, H., 105, 113, 114, 158, 166-168, 223
Speidel, D. H., 66, 223
Stewart, D. B., 128, 142, 223, 227

Tait, T., 35, 218
Tazief, H. K., 164
Thompson, R. N., 124, 153, 223
Tilley, C. E., 7, 8, 19, 29, 38-42, 59, 82-83, 101, 116, 134, 149-154, 163, 170, 223, 227
Tomkief, S., 113, 223

Traversa, C., 207
Turner, F. J., 12, 38, 54, 141, 169, 207, 223
Tuttle, O. F., 7, 12, 121-125, 129-130, 142, 144, 169, 206, 215, 221, 223, 224, 226
Tyler, R. C., 83, 224

Ulmer, G. C., 76, 224
Upton, B. G. J., 137, 227
Ussing, N. V., 158, 224

Van Zyl, J. P., 30, 224
Velde, D., 65, 107, 224, 227
Verhoogen, J., 12, 38, 54, 141, 164, 169, 207, 224
Verma, R. K., 172, 224
Vincent, E. A., 224
Vittaliano, C. J., 169, 224
Vogt, J. H. L., 6, 224
Von Eckerman, H., 144, 171, 224
Von Knorring, O., 217
Von Platten, H., 128-129, 224

Wadsworth, W. J., 134, 208
Wager, L. R., 26, 30, 82, 113, 224
Wagner, P. A., 171, 225
Walker, G. P., 113, 225
Wang, C. Y., 172, 183, 225
Washington, H., 98-99, 105-107, 225
Watkinson, D. H., 144, 225
Watson, E. B., 66, 72, 114, 218, 225
Wenlandt, R. F., 137-138, 225
White, J. L., 206
Wieblen, P. W., 113, 220
Wilkinson, J. F. G., 46, 225
Williams, A. J., 54, 220
Wilshire, H. G., 113, 225
Winkler, H. G. F., 225
Worst, B. G., 40, 225
Wright, F. E., 222
Wright, J. B., 54, 225
Wright, T. L., 40, 209
Wyllie, P. J., 7, 113, 144-147, 170, 183-185, 196, 199, 214, 215, 216, 225, 226

Yagi, K., 83, 226
Yoder, H. S., 6-9, 12, 19-20, 29, 35-39, 41-59, 62-65, 82, 90, 101, 103, 106-107, 113-120, 125-127, 136-137, 140, 142,

Author Index

144, 149-154, 163, 169-176, 180-185,
187-189, 193-196, 209, 215, 217, 221,
224, 226, 227
Yoshida, R., 62, 227

Yoshiki, B., 62, 227

Zavaritskii, A., 13, 41, 227
Zies, E. G., 164, 227

Subject Index

Adiabatic rise, 185
Agpaitic index, 162
  relation to melting interval, 162, 163-167
Agpaitic rocks, see Alkaline rocks
Albite melt:
  solubility of carbon dioxide in, 146
  solubility of water in, 143
Alkali basalt:
  chemical composition, 4, 170
  compositional continuity, 44
  compositional definition, 45
  degree of melting required for, 187-192
  depth of formation, 187-188, 190, 191-193
  depth of magma separation, 185, 192
  effect of volatiles on formation, 196
  melting relations, anhydrous, 156-158
  melting relations, hydrous, 151-152
  nodules (xenoliths) in, 171
  normative composition, 5, 45
  olivine and nepheline crystallization, 52, 63, 64
  solubility of water and carbon dioxide in, 143, 147
  volatile content in, 2
Alkali olivine basalt, see Alkali basalt
Alkali rock tetrahedron, 83
  compositional characteristics, 83-84
Alkali trend, 44
Alkaline rocks:
  agpaite-agpaitic index, 158, 162
  chemical composition, 4
  classification of, 85
  compositional plots of, 79, 84
  lava temperature, 164
  leucite-liquid reaction, 99-100
  leucite normative in, 85
  leucite stability in, 121, 123, 127, 131, 132
  liquid trends, 96-99, 105-107, 112
    peralkaline, 99
    potassic, 97-99, 105-107
    sodic, 52, 60, 64, 96-97
  melting relations:
    anhydrous-hydrous, 158-167
    under controlled P_{O2}, 167-168
  miaskitic, 168
  normative composition of, 5
  pseudoleucite formation, 99-100
  residual nature of, 77
  separation in space, 81-82
  silica saturation in, 78-80, 84-85
  solubility of water in, 143
  volatile content, melting interval, agpaitic index, 163-167
  volatile content of, 2
  volcanic gas composition, 2-3
Alkemade triangle, see Compositional plane
Amphibole, stability relations of, 149-152
Amphibolite, 152, 170
  relationship to basalt, 150-152
Ampholite, see Pyrolite
Analcite:
  in syenites, 132-133
  reaction relation of, 131-133
Anataxis, 125
Andesite:
  Cascade series, 69
  effect of P_{O2}, 67-76
  formation of, 72-75, 188-190, 200, 203
Anhydrous silicate systems, 8-115
Anorthosite, 29, 118
Aquous pore fluid in mantle, 147

235

Basalt:
  alkali, 45
  alkali trend, 44
  anhydrous melting, 154-156
  average chemical composition, 4, 170
  average mineral composition, 5
  classification normative, 45
  composition consistency with time, 194, 195
  composition of flood, 170
  crystallization-differentiation of, 13-15, 22-28, 51-54, 58-64
  definition, 8
  extrusion rates of flood, 194
  flood, 170, 194
  flow sheets, 52, 55, 63, 64
  glass, 170
  high alumina, 45
  lava temperature, 164
  major provinces, 194
  melting relations and melting intervals, 149-156
  relationship to source material, 170-178
  solubility of carbon dioxide, 146
  solubility of water, 145
  systems related to, 8-30, 34-76
  textures, 9, 17
  tholeiite, 19, 20, 43-45, 51-52, 170
  types of, 45
  univariant and invariant relations, *see* Flow sheets
  volatile content, 2
  volcanic gas composition, 2-3
  volume of flows, 194
  xenoliths in, 771
Basalt-andesite-rhyolite association, 44, 189, 190, 200, 203
Basalt tetrahedron:
  compositional characteristics, 42-44
  crystallization in, 51-54
  expanded, 54-55
  liquid trends (rock associations), 51-54, 64
  plane of critical silica undersaturation, 36-37, 42, 44
  plane of silica saturation, 42, 44
  simple, 42-43
  thermal divides in, 46, 54
Basaltic magma:
  ascent of, 184
  compositional consistency, 194-196
  depth-composition relations, 186-188
  depth-degree of melting, 187-193
  depth of separation, 188, 192
  as eutectic melts, 9, 52, 63, 195
  formation by complete or partial melting, 171
  heat for melting, 181-182
  low pressure changes in composition of, 157
  and low velocity zone, 179, 183-184
  pressure-temperature diagram, 149-155
  role of volatiles, 196
  seismic tracing, 186
  source materials:
    garnet peridotite, 172-175
    general characteristics, 169-171
    pyrolite, 175-178
  time required for melting, 182
Basanite:
  chemical composition, 4
  degree of melting, 189
  lava temperature, 164
  nepheline, 64
  normative composition, 5
  relationship to basalt, 52-64, 192
  solubility of water in, 143
  volatile content of, 2
Batch melting, 172
Binary systems, 8-12, 17-19, 117-121
Boundary curve:
  beginning of melting (solidus), 149, 151, 154, 156, 158, 159, 161, 175, 177
  cotectic, 13
  effect of pressure on, 189
  effect of volatile pressure on, 119, 135, 138
  four phase, *see* Univariant line
  fractionation, 22, 41, 161
  phase, 9
  reaction, 26, 78
  temperature direction, 9
Bushveld complex, South Africa, 40, 69

Calc-alkaline series, 44, 67-75, 118, 120, 189-190, 200
Carbon dioxide:
  effects on melting temperatures, 138-139
  estimates for upper mantle, 147

Subject Index

in lavas, 2
in rocks, 4
in volcanic gas, 3
solubility in mineral and rock melts, 146-147
Carbonatites:
  chemical composition, 5
  immiscibility, 113
Carnegie Institution, 6
  geophysical laboratory, 6, 140
Cascade series, 69
Chemical potential, 141
Chile-Peru region, depth of melting for, 184
Chromite reaction and composition, 39-40
CIPW norm and leucite, 85
Classification:
  of alkaline rocks, 85
  of basalts, 45
  of ultramafic rocks, 173
Columbia River plateau basalts, 194
  oxygen pressure effect, 167
Compatibility-incompatibility (antipathy) in minerals, 17-19, 59-60, 63, 65, 78
Component, 9
Composition:
  basalts (tholeiites), 170
  chemical and normative of igneous rocks, 4-5
  ranges in rock, 170
  ultramafic (peridotitic rocks), 174
  volcanic gases, 2-3
Compositional:
  plane, 33
  tie line, 14, 15
  triangle, 33
  zoning in feldspars, 12
Compositional controls:
  on crystallization, 22-26, 28-30, 52, 63, 69-75, 78-82
  on magma formation, 106, 186-193, 195
Congruent melting compound, 33
Continental crust, 203
Continental drift, see Plate tectonic magmatism
Coprecipitation:
  chromite-olivine, 24
  diopside-plagioclase, 9, 13-14
  forsterite-anorthite, 22
  principal phases, 28, 41
Cotectic boundary, 13
Cotectic fractionation paths, 24, 25
Crystal-liquid equilibrium, 15, 19
Crystallization:
  composition effect on, 19, 21
  equilibrium curves, 22, 41, 153
  fractionation curves, 22, 41, 161
  of melts, see various systems in Systems index
  pressure effect on, 117-118, 132-135, 138
  temperature effect on, 118, 119

Dacite, 44
Deccan plateau (trap) basalt, India:
  composition, 16
  rate of extrusion, 194
Degree of melting, 29, 185, 190, 192
Density:
  relationship of upper mantle source material, 172, 178, 179
  relationship to seismic velocities, 178
Depth-composition relation of magma, 106, 186-190, 192
Depth of magma separation, 184-185
Depth of melting, 186-193
Derivative magmas (residual magmas), 77
Diapiric uprise, 184-186
Diopside melt, carbon dioxide and water solubility in, 143, 146
Divariant surface, 30
Dunite, 26
  density, 178
  seismic velocity, 178

Eclogite:
  density, 178
  heat of melting, 181
  melting relations, 170
  principal minerals, 173
  relationship to basalt, 152-153
  seismic velocity, 178
  temperature-pressure relations, 152
  transition to gabbro, 153
Effect of P_{O_2}:
  on alkaline rocks, 167-168
  on basalt liquid trends, 66-75

Enstatite:
  density, 178
  incongruent melting of, 17
  relationship in forsterite bearing systems, 17, 20-21, 28, 42, 47, 74-76, 138
  seismic velocity, 178
Equilibrium:
  crystallization, 22
  curve, 22, 41, 153
  melting, 172, 195
Eudialite, 158-160
Eutectic:
  binary, 9, 18
  quaternary, 31, 33, 41, 55, 56, 63
  ternary, 21, 35, 46, 123
Experimental petrology, 3
  approaches in, 3-4, 148
  laboratories, 6-7
Exsolution:
  in feldspars, 120-121
  texture, 124

Feldspars:
  alkali, 120-122
  composition in rocks, 126
  crystallization, 11-13, 120-122, 125-130
  plagioclase, 119-120
  ternary, 125-130
  xenolith, 118
  zoning in, 120
Flood basalts:
  composition, 170
  major provinces, 194
  rate of extrusion, 194
Flow sheets:
  for basaltic trends, 52, 63, 64
  for mafic and felsic trend, 102-104
  for melilitic-potassic trend, 108
  for peralkaline trend, 111
  for potassic trends, 94, 96-98
  with univariant and invariant relations, 33
Fluid inclusions, 113, 140
Forsterite, see Olivine
Four component representation, 30-34
Four component systems, see Quaternary systems
Four phase curve, see Univariant line
Fractional crystallization, 23

Fractional melting, 172
Fractionation paths, 24-26

Gabbro, with layered intrusions, 26
Garnet peridotite:
  chemistry, 4, 174
  density, 175, 178
  heat of melting, 181
  heat production, 181
  mineralogy, 172, 173
  pressure temperature relation, 173-175
  seismic velocity, 178
Geophysical laboratory, 7, 140
Geotherm:
  oceanic, 177
  shield, 177
Granite:
  carbon dioxide solubility in melt, 147
  chemistry, 4
  compositional plot, 79
  eutectic composition, 9, 18
  heat production, 181
  hypersolvus, 124
  radioactive element abundance, 181
  subsolvus, 124
  water solubility in melt, 143
Granite minimum, 78, 82
Granite system, 78, 121-125
  eutectic in, 123, 132

Harzburgite, 173, 178, 179
Hawaiian basalt:
  chemistry, 4
  volatile content, 2
  volcanic gas, 2-3
Hawaiite, 44
Heat:
  of melting, 180
  of melting of basaltic compositions, 181
  production and melting, 182
  production rates, 181
  sources of, 180
Hematite-liquid reaction, 110-112
Henry's law, 142
Hornblende stability, see Amphibole
Hornblendite, 152, 170
Hydrous melting:
  of alkaline rocks, 158-167

Subject Index

of basalts, 149-153
of mantle, 189-193
Hypersthene basalt, 45, 52, 64

Ideal solution, 141
Igneous Petrology, objectives and problems, 1, 3
Igneous process, 1, 7
  major stages, 199
Igneous rocks, 3
  average bulk chemistry, 4
  normative mineralogy, 5
Ijolite point, 112
Ijolite-urtite association, 43, 112
Immiscibility, liquid, 113
  causes of, 113-114
Incongruent melting component, 17, 27, 108-109
Invariant point, 17
  quaternary, 33-34, 39, 52, 53, 56-57, 62, 66, 68, 74, 92-96, 104, 108, 111
Iron enrichment:
  index, 69-70
  relationship with liquidus temperature, 153, 163, 165
  relationship to oxygen pressure, 72, 153, 167, 168
  relationship with silica, 69-73
  trends, 66-69
Island arcs, 189, 190, 202, 203

Joins (planes):
  alkali feldspar, 78
  in Alkali Rock Tetrahedron, 83, 86
  in Basalt Tetrahedron, 42, 44, 55, 56-61
  compositional, 33
  diopside-albite, 48
  forsterite-albite, 46, 47
  in iron-bearing quaternary system, 33, 70-75, 108-112

Karnataka, India, 76
Karoo lava, 194
Kimberlite, 171, 192, 196
  xenolithsin, 171

Lava:
  extrusion rates, 194
  temperatures, 164

Layered intrusions, rock associations in, 26, 29, 30, 76
Lever rule, 10
  application, 10, 12
Lherzolites, 171, 173
Liquid fractionation, 113
Liquid immiscibility, 113-114
Liquid trends:
  alkaline magma, 94, 98, 104-108
  mafic magma, 52, 64, 104-109
  peralkaline, 99, 110-112
Liquidus, 11
Liquidus surface, 13
Lithospheric plate, 202, 204
Low melting compositions, *see* Eutectic; Invariant point
Low velocity zone:
  compositional features, 179, 189
  degree of melting, 183-184
  incipient melting, 183-184

Magma:
  ascent, 184
  ascent rate, 186
  calc-alkaline, 188-190
  compositional consistency of basalt, 193-194
  definition, 1
  degree of melting, 29, 185, 188-190, 192
  depth-composition relation, 186-190, 191-193
  depth of separation, 184
  role of volatile, 196
Magmafracting, 185
Magma mixing, 29
Magmatism-mantle differentiation, 202-204
Mantle composition, 172-179, 203
Megabar, 7
Melilite leucitite, 108
Melilite nephelinites, 62-64
Melilite-olivine nephelinites, 62-64
Melilite-plagioclase incompatibility, 57, 59, 60, 65
Melilitites-olivine, 62-64
Melting:
  batch, 172
  beginning of, 150, 151, 155, 156, 159-161. *See also* Solidus

crustal, 124
curves (liquidus curves), 10, 11
  fractional, 172
  interval, 149, 154, 162, 163-167
  invariant, 195
  partial, 172
Mid oceanic ridges, 194, 203
Mixing of melts, 29
Mugearite, 44

Nepheline melt:
  solubility of carbon dioxide, 146
  solubility of water, 143
Nepheline basanite, 45, 52, 54, 62-64, 190, 192
Nephelinite:
  melilite, 62-64
  olivine, 45, 52, 62-64
  solubility of carbon dioxide, 146
  solubility of water, 143
Nepheline syenite:
  agpaitic, 158
  chemistry, 4
  melting relations, 158-168
  mineralogy, 5
  solubility of water, 143
  system, 79
  volatile content, 2
Nockold's Averages, 5, 69
Nodules, 171
Non-equilibrium, 12
Norite (quartz), 26
Normative classification:
  alkaline rocks, 85
  basalts, 45

Ocean floor basalt, volatile content, 2
Oceanic:
  basalt (tholeiite), 170
  crust, 179
  geotherm, 177
  lithosphere, 202, 203
Olivine control:
  in magma composition, 187-188
  in melt composition, 22-28
Olivine-liquid reaction, 18-21, 52, 63
  complete, 19-20
  none, 19-20
  partial, 19-20

Olivine melilitite, 64, 190, 192
  solubility of carbon dioxide, 146
  solubility of water, 143
Olivine melt:
  solubility of carbon dioxide, 146
  solubility of water, 143
Olivine-quartz antipathy, 19, 21
Orogenic association, *see* Calc-alkali rocks
Oxygen partial pressure, 66-75, 167-168

Palisade sill, 69
Parent magma, 169
Paricutin, 69
Partial melting, 172
  initial and eventual, 185
Peralkaline trend, 99
Peridotite types, 173-175
  heat production of, 181
  radioactive elements in, 181
Peritectic:
  in binary systems, 18, 19
  in quaternary systems, 33, 63, 91-95, 101-103
  in ternary systems, 21, 36, 37, 47, 48, 78-80
Petrogeny's residua system:
  anhydrous relations, 77-82
  hydrous relations, 130-134
Petrological:
  diversity, 1
  objectives, 1-3
  research, 1
Phase:
  areas, 30
  volumes, 30-31
Phonolite system, 79
Picrite basalt, 182, 187
Piercing points, 37, 53, 62, 90, 103
Plagioclase feldspar:
  anhydrous relations, 11-12
  composition zoning, 12, 15, 119
    normal, 12, 15, 120
    oscillatory, 120
    reverse, 120
  hydrous relations, 119-120
  incompatibility with melilite, 57, 59, 65
Plate movement-incipient melting, 183-184
Plate tectonic-magmatism, 202-204
Poisson's ratio, 171

Subject Index

Pore fluid aquous, 147
Precambrian shield geotherm, 177
Pseudoternary system, 21
Pyrolite:
  chemistry, 174
  composition, 176
  density, 178
  melting relations, 191
  pressure-temperature relations, 176-177, 191
  radioactive element abundance, 181
  seismic velocity, 178
  types, 176
Pyroxenite, 171, 178

Quartz tholeiite, 19, 20, 43, 45, 51-52, 190
Quaternary representation, 30-34
Quaternary subsystems, 33, 42, 43, 53, 62, 84, 93, 104
Quaternary systems, 30, 34, 41, 55, 66, 69, 72, 101, 108, 114
  with eutectic, 31, 33, 41, 55, 63, 68, 72, 110, 111
  with minimum, 86, 89, 95, 100
  with peritectic, 33-34, 39, 41, 55, 63, 67, 68, 74, 100, 108, 111

Radioactive:
  element abundances, 181
  heat production, 181
Reaction curve, 26
Reaction point, *see* Peritectic
Reaction series (principle), 6, 152
Residual liquids (magmas), 77
Rhyolite, 44, 79, 84

Salic contamination, 29
Seismic velocity:
  in rocks and minerals, 178
  in upper mantle, 178
Silica:
  enrichment trend, 44
  oversaturation, 19, 27-28, 42-44, 78
  undersaturation, 19, 42-44, 78-79
Skaergaard, Greenland, 69
Solid solution, 11, 118, 121
Solidus:
  alkaline rocks, 159-161

  basalt, 150, 151, 154, 156
  garnet peridotite, 175
  pyrolite, 177, 191
Solubility:
  of carbon dioxide in melts, 146-147
  expression, 141-142
  of water in melts, 143
Sources:
  of heat for melting, 181
  of volatiles in mantle, 147, 196
Spinel reaction, 24-25, 38-40
Stillwater complex, 69
Subsolvus granites, 141

Tachylite, 170
Tephrite, 52, 54, 64
Ternary systems (and joins), 8, 13, 20, 45, 55-61, 77
  with eutectic, 13, 21, 35, 46, 128, 195
  with no eutectic but minimum, 13, 36, 37, 46, 50, 87, 123, 127, 132
  with peritectic, 21, 35, 46, 77, 86, 88, 131, 132
  with thermal divides, 46, 48, 70
Tetrahedron:
  alkali rock, 82
  basalt, 42
  expanded basalt, 54-55
Textures:
  eutectic (ophitic), 9, 124
  graphic, 124
  hypidiomorphic, 124
  ophitic, 9, 17
  porphyritic (phenocryst), 168
Thermal valley, 79, 82
Tholeiite:
  chemistry, 4, 170
  mineralogy, 5
  olivine, 43-45
  quartz, 43-45
  relations to other basalts, 43, 52, 63, 64, 189, 192
  solubility of carbon dioxide, 147
  solubility of water, 143
  see also Basalt
Three component systems, *see* Ternary systems
Trachyte, 44, 84
  leucite, 84-85, 94, 98, 104

quartz, 84
Two component systems, *see* Binary systems
Type of basalts, 45

Ultramafic rocks:
  chemistry, 174
  classification, 173
  mineralogy, 173
  seismic velocities, 178
Undepleted source, 172
Univariant and invariant relations, *see* Flow sheets
Univariant line, 30, 37, 52, 53, 62, 63, 90, 92, 93, 94, 103, 108
Upper mantle:
  compositional nature, 171, 175-176, 179, 203
  density, 178
  Poissons ratio of, 171
  seismic velocities, 171, 178

Volatile contents, lavas and rocks, 2

Volatile effects, 115-116
  in magma generation, 196
Volatile solubility in silicate melts, 143, 146-147
Volcanic gases, 2-3

Water content:
  in lavas, 2
  in rocks, 4
Water solubility in silicate melts, 143
Water solubility mechanism, 115
Water vapor pressure effects, 115-139, 149-169

Xenoliths:
  in basalts, 171
  in calc-alkaline rocks, 118
  garnet peridotitic, 171
    chemistry, 171
    mineralogy, 174
  in kimberlites, 171

Zoning in feldspars, 12, 14, 118-120

Systems Index

MINERAL SYSTEMS WITHOUT WATER

Albite-Anorthite, 10
Albite-Orthoclase, 121
Diopside-Anorthite, 8
Forsterite-Silica, 17
Sphene-Anorthite, 9

Albite-Akermanite-Diopside, 60
Albite-Anorthite-Orthoclase, 125, 127
Albite-Orthoclase-Silica, 78
Albite-Wollastonite-Akermanite, 57
Diopside-Albite-Anorthite, 13
Diopside-Albite-Enstatite, 51
Diopside-Albite-Forsterite, 36
Diopside-Albite-Leucite, 87
Diopside-Albite-Nepheline, 48
Diopside-Albite-Orthoclase, 86
Diopside-Albite-Silica, 48
Diopside-Enstatite-Silica, 20, 50
Diopside-Forsterite-Anorthite, 35
Diopside-Forsterite-Enstatite, 20, 50
Diopside-Forsterite-Nepheline, 49
Diopside-Forsterite-Pyrope, 195
Diopside-Forsterite-Silica, 20, 50
Diopside-Leucite-Silica, 27
Diopside-Nepheline-Akermanite, 56
Diopside-Nepheline-Sanidine (Orthoclase), 88
Diopside-Nepheline-Silica, 48
Forsterite-Albite-Anorthite, 36
Forsterite-Albite-Enstatite, 47
Forsterite-Albite-Nepheline, 20, 47
Forsterite-Albite-Silica, 20, 47
Forsterite-Anorthite-Silica, 20, 21
Forsterite-Leucite-Silica, 28
Forsterite-Nepheline-Silica, 47, 189
$MgO\text{-}Fe_3O_4SiO_2$, 71
Nepheline-Akermanite-Albite, 59
Nepheline-Akermanite-Wollastonite, 58
Nepheline-Kalsilite-Silica, 78
Nepheline-Wollastonite-Diopside, 61

A-B-C-D, 30-34
Diopside-Albite-Anorthite-Ferrosilite, 15
Diopside-Enstatite-Albite-Silica, 42, 52
Diopside-Forsterite-Akermanite-Leucite, 107
Diospide-Forsterite-Albite-Anorthite, 34, 39
Diopside-Forsterite-Albite-Enstatite, 42, 52
Diopside-Forsterite-Nepheline-Albite, 42, 52
Diopside-Forsterite-Nepheline-Silica, 41
Diopside-Nepheline-Kalsilite-Silica, 82
Forsterite-Anorthite-Chromite-Silica, 24
Forsterite-Anorthite-Magnetite-Silica, 72
Forsterite-Anorthite-Orthoclase-Silica, 29
Forsterite-Nepheline-Larnite-Silica, 54
Forsterite-Wollastonite-Iron oxide-Silica, 67
Forsterite-Wollastonite-Magnetite-Silica, 68
$MgO\text{-}FeO\text{-}Fe_2O_3\text{-}SiO_2$, 69
$Na_2O\text{-}Fe_2O_3\text{-}Al_2O_3\text{-}SiO_2$, 108

Diopside-Forsterite-Anorthite-$MgCr_2O_4$-Silica, 24, 40
Diopside-Forsterite-Nepheline-Albite-Leucite, 100
Forsterite-Fayalite-Anorthite-Orthoclase-Silica, 114
Orthoclase-Albite-$FeO\text{-}Fe_2O_3\text{-}SiO_2$, 114
$K_2O\text{-}CaO\text{-}FeO\text{-}Fe_2O_3\text{-}Al_2O_3\text{-}SiO_2$, 114

MINERAL SYSTEMS WITH WATER AND CARBON DIOXIDE

Albite-Anorthite-H_2O, 118
Albite-Orthoclase-H_2O, 120
Diopside-Anorthite-H_2O, 117
Diopside-Sanidine-H_2O, 137

Albite-Anorthite-Orthoclase-H_2O, 125

Albite-Orthoclase-Silica-H_2O, 122
Forsterite-Kalsilite-Silica, H_2O, 137
Forsterite-Kalsilite-Silica, H_2O+CO_2', 137
Forsterite-Nepheline-Silica, H_2O, 138
Nepheline-Kalsilite-Silica-H_2O, 130

Albite-Anorthite-Orthoclase-Silica-H_2O, 128
Diospide-Nepheline-Albite-Acmite-H_2O, 136
Forsterite-Albite-Anorthite-Silica-H_2O, 134
Forsterite-Albite-Anorthite-Silica-H_2O, 135
    (with 10 wt% orthoclase, 135)

K_2O-MgO-Al_2O_3-SiO_2-H_2O-CO_2, 138

WHOLE ROCK SYSTEMS

Alkali basalt, 153, 156
Alkali basalt + H_2O, 149, 151
Lujavrite, 160
Lujavrite + H_2O, 160
Naujaite, 159
Naujaite + H_2O, 159
Nepheline syenite, 160
Nepheline syenite + H_2O, 160
Olivine tholeiite, 153, 155
Olivine tholeiite + H_2O, 149, 150
Peridotite, 173, 175
Phonolite, 161
Phonolite + H_2O, 161
Pyrolite, 175, 177